Applied Spectroscopy and the Science of Nanomaterials

Applied Spectroscopy and the Science of Nanomaterials

Editor

Roushan Singh

Applied Spectroscopy and the Science of Nanomaterials

Edited by **Roushan Singh**

Printed in 2017

ISBN: 978-1-68117-210-1

Library of Congress Control Number: 2015936573

© 2016 by
SCITUS Academics LLC,
616, Corporate Way, Suite 2, 4766,
Valley Cottage, NY 10989

www.scitusacademics.com

Contents

vi

Preface

Applied Spectroscopy and the Science of Nanomaterials is a synergistic compendium of informative and cutting-edge chapters written by leading researchers in the fields of spectroscopy and condensed matter physics as applied to a variety of materials at the nanoscale. Applied spectroscopy is the application of various spectroscopic methods for detection and identification of different elements/compounds in solving problems in the fields of forensics, medicine, oil industry, atmospheric chemistry, pharmacology, etc. Nanotechnology is based on the fact that nanostructures, nanodevices, and nanosystems exhibit novel properties and functions as a result of their small size. It is a highly multidisciplinary field, drawing from such subjects as applied physics, materials science, colloidal science, device physics, supramolecular chemistry, and mechanical and electrical engineering.

Editor

Synthesis Optimisation of Lysozyme Monolayer-Coated Silver Nanoparticles in Aqueous Solution

A. V. Yakovlev and O. Yu. Golubeva

Institute of Silicate Chemistry of Russian Academy of Sciences, Admirala Makarova Emb. 2, Saint Petersburg 199034, Russia

ABSTRACT

This paper presents an optimisation of the synthesis of silver nanoparticles encapsulated in a biological shell. The synthesis was carried out in an aqueous solution of silver nitrate. Sodium borohydride was used as a reducing agent. Lysozyme served as a bioactive coating agent. The samples produced were studied using dynamic light scattering, transmission electron microscopy, and UV-Vis spectroscopy. The function of the dependence of the reagent ratio in obtained sols on optical properties is shown. Furthermore, the influence of the synthesis temperature, reactant ratio, and order of mixing on the particle size distribution parameters is shown. The optimal reagent mass ratio, $NaBH_4 : LYZ : AgNO_3 = 0.22 : 0.77 : 1$, is established. The resulting composition allows the synthesis of particles with a mean diameter of 18 nm and a bioshell thickness of ≈ 3.5 nm. Moreover, the necessity of the synthesis optimisation and precise parameter control is clearly demonstrated.

INTRODUCTION

During the past few decades, many researchers [1] from various fields of science have been involved in the synthesis of noble metal nanoparticles for physics, chemistry, biology, and medicine applications [2]. This occurred because of the unique physicochemical properties of such materials, which make it possible to solve a wide range of problems in medicine, biology, catalysis, optics, and so forth [3–9].

The use of silver as an antibacterial agent is well known [10, 11]. However, ionic silver exhibits high toxicity properties to human cells [11] and, at present, its application in the ionic form is extremely restricted. One of the ways to solve the usability issue with silver is to utilise silver in nanoparticles, rather than in the ionic form. This method not only avoids the toxicity concerns but also yields unique material properties, such as providing a selective action on microorganisms through bioactive agents by targeted delivery on the surface of carrier nanoparticles [12]. Chemical and photochemical reductions in solutions are the simplest and most common methods of silver nanoparticle synthesis. These methods allow for the generation of small nanoparticles with a definite shape [13]. Furthermore, the use of composites based nanoparticle is common and has a good potential in everyday use [14]. However, the issue of the instability of the obtained sols has not been completely solved. Moreover, for medical applications, the commonly used stabilisers may be very toxic or allergenic for humans. Furthermore, strongly diluted solutions are used in conventional synthesis techniques for the production of particles smaller than 50 nm. This results in a low quantity of nanoparticles [15] in solutions, which makes the study of such objects' properties and their practical application difficult. Therefore, in this paper, a more promising particle stabilisation technique, which is directly related to particle synthesis optimisation and functionalisation by creating bioactive protective shells, will be described. In this case, the obtained particles have a metal core nanoparticle and a biomolecular shell. The synthesis of these substances (bioconjugates) removes some of the problems that arise in the synthesis of nanoparticles in solutions, for example, the issues of stability and nanoparticle dispersion. As demonstrated by Eby et al. [16], such ferments as lysozyme can not only build up strong shells on the nanoparticle surface but also serve as a catalyst in the photochemical synthesis of silver nanoparticles. Despite the simplicity of this technique, it has certain shortcomings and has clear disadvantages for industrial implementation [7]. The main difficulty stems from using noxious and/or organic media for the production of nanoparticles. The purification of such sols is a very time-and-energy-consuming process.

The most critical parameter for the use of bioconjugates is their size [17–19]. The presence of even a few particles larger than 60 nm can make it impossible to predict the biological effect of bioconjugates or to compare the results of their study. Commonly, to reduce the negative effects of particle size differences and to achieve a nearly monomodal size distribution, various complex purification methods are used [20, 21]. As mentioned previously, these methods are laborious and greatly reduce the amount of nanoparticles in sols. Some investigators use different techniques to control particle size, such as micellar synthesis or stabilisation in matrices [22–24].

The proposed synthesis method is based on the formation of protein lysozyme monolayer-coated silver nanoparticles in an aqueous solution of a three-component synthesis system of source substances ($LYZ : NaBH_4 : AgNO_3$).

Our main objective is the study of the basic synthesis regularities and the selection of the optimal synthesis conditions, which has not been reported previously [25–27]. The selected conditions should produce silver nanoparticles smaller than 50 nm in size, with a narrow particle size distribution.

Synthesis using the suggested conditions involves certain shortcomings and problems to be solved. As described previously, silver nitrate interacts with lysozyme in sunlight. This interaction results in the production of nanoparticles coated with a lysozyme shell. The size of the particles produced in the aqueous solution is 60–80 nm, and the particle size distribution may differ considerably from sample to sample. For the proposed method, this side effect is a negative contribution to the particle size distribution that should be avoided or accounted for in the experiments. Because the reaction proceeds in an aqueous medium, it is impossible to completely exclude the solvent effect [28]. The experiment is also complicated by the concurrent sodium borohydride hydrolysis reaction. In practice, excess sodium borohydride is used to reduce the negative effect of hydrolysis. For many years, this procedure has been well-proven. However, this method shows a significant disadvantage: the use of excess sodium borohydride causes an increase in the concentration of its hydrolysis products ($NaBO_2$), which may strongly affect the long-term stability of the system. Therefore, it is necessary to carry out a study of the optimal weight ratio of the initial reagents that allows stable hydrosols production with the required physicochemical parameters while maintaining long-term stability. Despite the disadvantages mentioned above, we assume that the considered synthesis system is convenient for a study of the nanoparticle biocoating phenomena in question and can be adapted for practical use in medicine.

We attempt to demonstrate the importance of an in-depth investigation of the processes and conditions of nanoparticle formation and substantiate the effect of precise synthesis performance on the basic physicochemical parameters of the lysozyme-coated nanoparticles. The data obtained will allow the acquired knowledge to be extended to other systems and will provide researchers with a more convenient, practical, and cost-effective method of nanoparticle synthesis.

MATERIALS AND METHODS

Reagents

All the reagents, hen egg lysozyme (LYZ, ≈70000 units/mg, Sigma-Aldrich, USA), sodium borohydride ($NaBH_4$, 99%, Sigma-Aldrich, USA), and silver nitrate ($AgNO_3$, 99.9%, Khimmedsintez, Russia), were of a high purity grade and were used without additional purification. To prepare the solutions, we used deionised water produced from distilled water using a Vodolei setup (NPP Khimelektronika, Russia) with specific conductivity no higher than 0.2 µS/cm. The experiments were performed in sterile, disposable 2 mL microvials.

Methods of Analysis and Instrumentation

To prepare the stock solutions, volumetric flasks of first-class accuracy, pretreated with a chromic mixture and washed no less than ten times with deionised water, were used. The temperature was controlled in a CH-100 thermostat (Biosan, Latvia) with holes for the microvials and a built-in Peltier element that set the temperature range at (−20–100) ± 0.1°C. The reagents were mixed on a V-1 vortex (Biosan, Latvia). Particle sizes were measured with the dynamic light scattering method using a Nanotrac Ultra instrument (Microtrac, USA) with no special sample preparation. The particle size distribution was computed using the Microtrac Flex 10.6.1 software (Microtrac, USA). Additionally, transmission electron microscopy (TEM) was used to measure the particle sizes on a Jem 2100-F microscope (JEOL, USA) at U_{ac} up to 200 kV. The samples were prepared using a standard carbon replica technique.

Micrographs were processed using the Gimp 2.8 graphics editor. The particle size distribution graph was constructed and approximated by the Gauss method using the Python 2.7.5 programming language and the SciPy library SciPy [29]. A spectroscopic study was carried out on a UV-Vis UV-2600 spectrophotometer (Shimadzu, Germany) using cells with an optical path length of 5 mm. Before measurements, the samples were diluted 10-fold, and the total absorption spectra were measured. For all the samples, the position of the absorbance peak and its area were determined.

The postsynthesis content of the residual ionic silver was measured with ion-selective analysis using an Ekspert-001 instrument (Ekoniks-Ekspert, Russia).

Nanoparticle Synthesis

The nanoparticles were synthesised in aqueous solutions using a chemical reduction of silver nitrate by sodium borohydride in the presence of lysozyme.

Solutions with the following concentrations were prepared for experiments: $C_{AgNO_3} = 10$ mM, $C_{NaBH_4} = 20$ mM, and $C_{LYZ} = 3$ mg/mL. A fresh sodium borohydride solution was prepared for each series of experiments, and it was immediately cooled to a temperature of $0°C$. The necessary amounts of the sodium borohydride and silver nitrate solutions were transferred to the microvials by an autopipette. Then, the lysozyme solution was added (half of the required amount to the silver nitrate solution and the other half to the sodium borohydride solution) to obtain the desired $NaBH_4 : LYZ : AgNO_3$ weight ratio. The lysozyme and silver nitrate solution mixture was shielded from sunlight. The reagent mixing weight ratio was varied in such a way that the amount of reduced silver and the final volume of the solution remained constant. This was achieved by adding the required volume of deionised water.

Before mixing, the solutions were cooled or heated to the required temperature in the thermostat for ten minutes to complete temperature stabilisation. Then, the solutions were mixed and, after synthesis, naturally equilibrated to room temperature. Obtained samples were analysed using the previously described methods.

RESULTS AND DISCUSSION

It was shown that the results' reproducibility, sol stability, and single-mode particle size distribution could only be obtained with the proper order of reagent mixing as described above. All other methods of solution preparation resulted in a bimodal particle size distribution or a very broad particle size distribution. Moreover, the solution mixture stirring rate did not play any part within the tested range of 700–3500 rpm.

Let us consider the effect of the $NaBH_4 : AgNO_3$ weight ratio on the sol parameters. According to the experimental data, an excess of sodium borohydride resulted in instantaneous particle precipitation. Therefore, our study was conducted within the weight ratio $NaBH_4 : AgNO_3 < 1$ g/g. An excess of silver nitrate up to the weight ratio $NaBH_4 : AgNO_3 = 0.1$ g/g did not significantly affect the particles size or their distribution. However, when exposed to light, the excess silver nitrate reduced to form particles 80 to 100 nm in size, which affected the general particle size distribution. Therefore, for the next study, we chose the ratio $NaBH_4 : AgNO_3 = 1$ g/g. At this precursor-to-reducing-agent ratio, the residual amount of silver ions does not exceed a concentration of 10^{-6} mol•L^{-1} in all the sample series.

Let us consider the effect of the $LYZ : AgNO_3$ ratio depicted in Figure 1. The experiments were carried out at the equivalent $NaBH_4 : AgNO_3$ ratio and a solution temperature of 0°C. The obtained dependence can be broken up into the following three main weight ratio ranges.

- The 0.0–0.6 g/g range corresponding to lysozyme deficiency. The nanoparticles are stabilised at the initial growth stage. The uncoated particles undergo uncontrolled growth and aggregation phases.

- The 0.6–1.0 g/g range corresponding to the stability condition with the optimum reagent ratio. Within this range, a complete lysozyme monolayer coating of the particles occurs.

- The range >1.0 g/g corresponding to particle conjugation completion and the onset of secondary lysozyme layer formation over the monolayer. This process results in an increase in the hydrodynamic radius, which is revealed by the dynamic light scattering method and by a change in the optical properties of the particles.

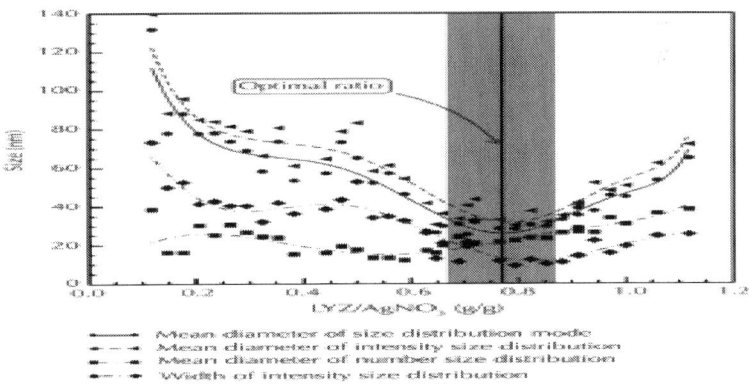

Figure 1: The effect of lysozyme-silver nitrate weight ratio on the parameters of the particle size distribution, obtained by dynamic light scattering.

To better understand and interpret the data, let us consider each dependency individually. As follows from Figure 1, the curves representing the mean particle diameters corresponding to the major mode of the particle size distribution determined by the measured light scattering intensity and to the total particle size distribution virtually coincide and exhibit similar behaviour, which unambiguously excludes the bimodal nature of the particle size distribution. As mentioned earlier, the particle size is of vital importance, even if the number of such particles is insignificant.

It should be noted that the width of the intensity size distribution is similar to that of the curves discussed above, and their minimum lies at the ratio LYZ : $AgNO_3 \approx 0.77$ g/g. The total distribution width at this region is less than 20 nm. Based on the importance of this parameter, it would be erroneous to consider only the mean particle diameter of the size distribution by number. In this case, the application of the dynamic light scattering technique has a great advantage because the larger particles can scatter light much better and contribute more to the distribution by intensity than smaller particles. Through the examination of the number distribution of the particle sizes shown in Figure 1, it can be observed that the values for all the ratios are similar, with a slight bend at the weight ratio of 0.6 g/g and an increase after a weight ratio of 0.9 g/g. This distribution is closest to the actual distribution, which can be obtained by transmission electron microscopy without taking into account the difference between the hydrodynamic diameter of the particles with an organic shell and the actual diameter of the core-nanoparticles observed by transmission electron microscopy images. From the data obtained, the best stabilisation and optimal consumption of reagents in conjunction with the completely homogeneous properties of the obtained products occur in a narrow interval of reactant ratios, with a weight ratio of approximately $NaBH_4$: LYZ : $AgNO_3$ = 0.22 : 0.77 : 1. The resulting value is in good agreement with data obtained by Eby and others but has greater accuracy [16].

In addition to dynamic light scattering, UV-Vis spectroscopy provides very important information about the optical properties of the obtained bioconjugates. The spectrum presented in Figure 2 clearly exhibits the absorption band with a peak at a wavelength of 404 nm.

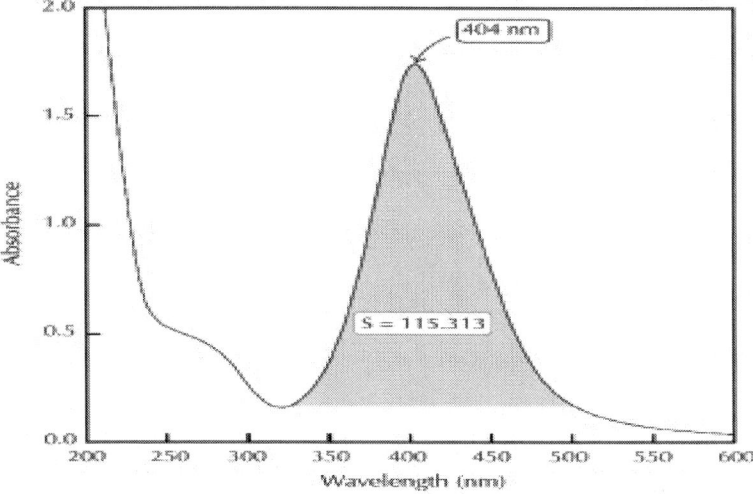

Figure 2: UV-Vis absorption spectrum of the bioconjugate sol with optimal parameters of synthesis: $NaBH_4 : LYZ : AgNO_3 = 0.22 : 0.77 : 1$ and t=0°C. S—area under the curve.

It is known that, for such small particles, the absorption wavelength should be in the range of 398–400 nm [30]. For the bioconjugates, the absorption wavelength shift may be evidence of biomolecular interaction with the surface of a nanoparticle.

The absorption spectrum may be described based on the following three main characteristics:

- the position of the peak associated with the particle size,
- the absorbance intensity directly related to the number of nanoparticles,
- the peak area that can implicitly indicate the presence of larger nanoparticles.

In Figure 3, these parameters are presented as a function of the reagent ratio.

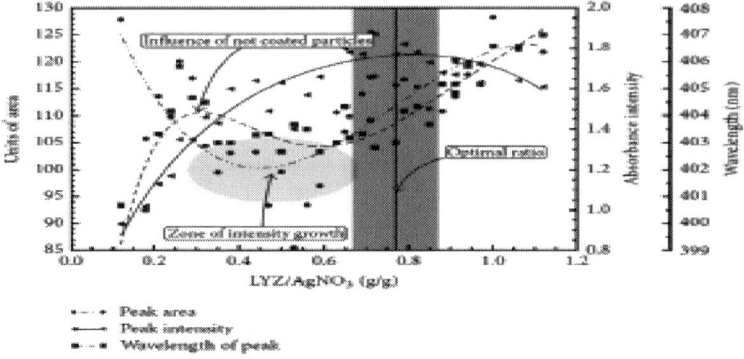

Figure 3: Absorbance peak area, its intensity, and the wavelength corresponding to the peak position as functions of the $NaBH_4 : LYZ : AgNO_3 = 0.22 : 0.77 : 1$ and t=0°C.

It is worth mentioning that the visual colour of the solution on the orientation, in the case of systems having a sufficiently high concentration of nanoparticles, has no practical utility and can be wrongly interpreted. During synthesis, the samples immediately became dark yellow in colour and nearly opaque after adding the reducing agent. Therefore, determination of the "good" samples from "bad" in this manner is not possible. However, when the sol is diluted, a characteristic yellow colour appears.

If we consider that the amount of reduced silver ions in the solutions was kept constant and equal to the molar concentration of sodium borohydride solution, the intensity of absorption is fully responsible for the amount of nanoparticles that exhibit plasmon resonance.

The curves shown in Figure 3 describe the change of the peak area, the intensity, and the wavelength of the peak maximum. Maximum intensity undergoes changes stronger than all the other parameters. The peak of this curve is located in the area-to-mass ratio of lysozyme silver nitrate ≈0.77 g/g, which corresponds to the optimal weight ratio obtained by dynamic light scattering. This finding indicates that the most efficient light absorbing sols are in the region where the optimal weight ratio is.

The curve of the peak wavelengths has a more intricate shape and cannot be easily explained. Referring again to Figure 1, we can conclude that with weight ratios less than 0.4 g/g there is a slow decrease in particles size and they begin to absorb light. Additionally, the particle size is large due to the incomplete shell, and the wavelength of maximum absorption is shifted to the red region of the spectrum (which may be caused by influence of uncoated particles). In this region of weight ratios, the results may fluctuate strongly in the experiments because uncontrolled particle aggregation occurs, and the particle size distribution is far from optimum. When the weight ratio is increased beyond 0.4 g/g, the influence of conjugated nanoparticles is noticeable.

At weight ratios of approximately 0.77 g/g, complete coverage of the particles occurs. The wavelength of the peak shifts slightly due to the interaction of lysozyme with the surface of the particle. Increasing the amount of lysozyme further leads to a multilayer coating and a change in the optical properties of the particles. However, it should be noted that the variation in the wavelength corresponding to the absorbance peak position is very small and occurs within 399–407 nm.

The curve of the peak area in Figure 3 definitively confirms previous findings. At weight ratios less than 0.4 g/g, the peak area is formed due to the peak broadening caused by the presence of large-sized particles. At a weight ratio of ≈0.5 g/g, the peak narrows and then its area only increases due to the increasing absorbance intensity (zone of intensity growth in Figure 3).

To explain the obtained result more easily, let us examine Figure 4. It is clear that, due to the lack of lysozyme in the system, the absorption spectrum has a broad base and relatively weak intensity. The broadening is also noticeable in the range of the high $LYZ:AgNO_3$ weight ratio. The peak of this curve, situated in the 0.6–0.8 g/g weight ratio range, is where the peak base is the narrowest and has its maximum intensity.

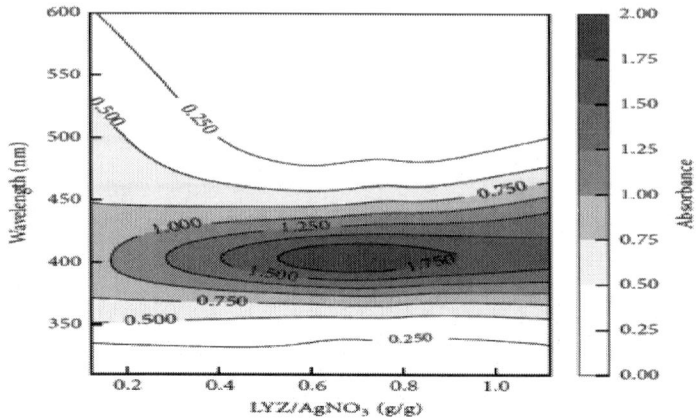

Figure 4: Contour plot displaying changes in the nature of the UV-Vis transmission spectra curve as a function of the $LYZ:AgNO_3$ weight ratio and at t=0°C.

Figure 5 shows a micrograph of a sample prepared at the optimum reagent ratio. This micrograph shows that the particles are spaced apart, and no large aggregations are observed. A particle with a regular spherical shape is observed in the higher-resolution micrograph in the inset in Figure 5. However, it appears impossible to determine the parameters of the biological shell from the micrographs because of its small size and poor contrast. But on the micrograph (Figure 6) of sol, obtained in lack of lysozymes which have larger size of nanoparticles, the shell is easier to capture and measure. From the data obtained, the size of the shell is ≈3.7 nm.

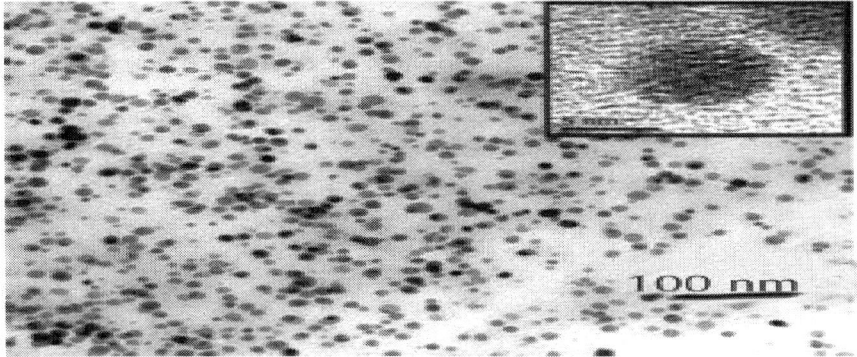

Figure 5: Micrograph of a sample synthesised in optimal parameters of synthesis at the weight ratio $NaBH_4 : LYZ : AgNO_3 = 0.22 : 0.77 : 1$, t=0°C.

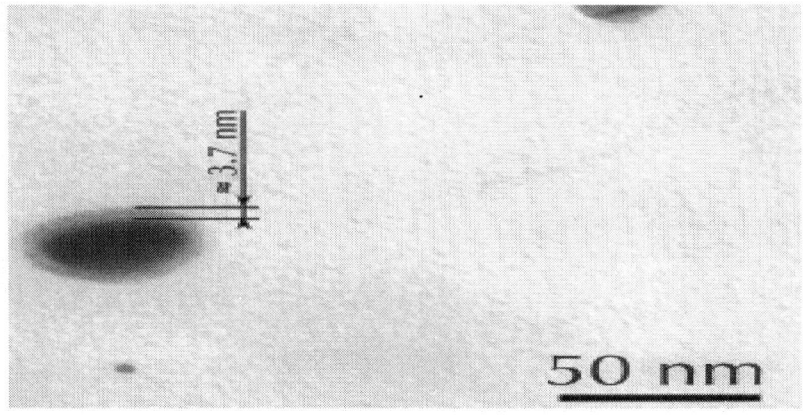

Figure 6: Micrograph of a sample synthesised in nonoptimal weight ratio $NaBH_4 : LYZ : AgNO_3 = 0.22 : 0.6 : 1$, t=0°C.

The comparison of particle size distributions in Figure 7(a), obtained by different methods (DLS and TEM), shows that certain curves shift relative to each other.

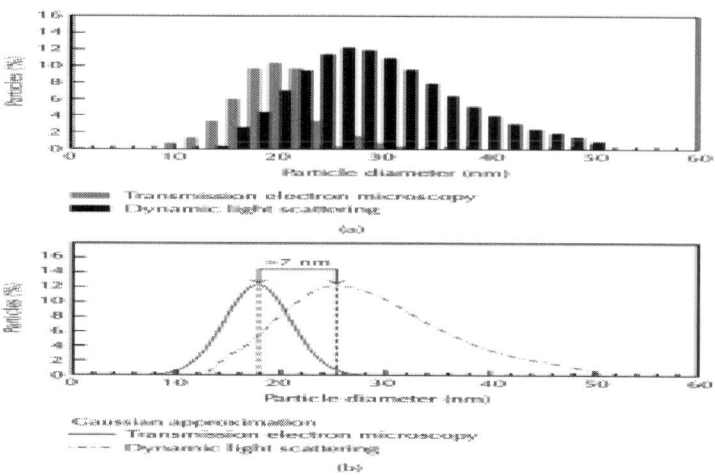

Figure 7: Comparison of the particle size distributions obtained by the dynamic light scattering and TEM methods (analysis of more than 800 particles). (a) Actual distributions; (b) Gaussian approximation with the distribution mode indicated.

Shift values between the curves at Figure 7(b) reflect the difference between the hydrodynamic diameter of the bioconjugate (d=25.1nm and SD=11.2nm) and the size of the metallic core-nanoparticles (d=17.0nm and SD=7.3nm); this difference is ≈7 nm. Consequently, the thickness of the coating on the particle surface is ≈3.5 nm, considering that the error of the methods matches the lysozyme hydrodynamic size, which should be ≈2–5 nm [28], depending on the properties of the solution, the location of a molecule on the surface of its structure, and other physicochemical properties. Furthermore, conformational changes in the lysozyme structure [31–34] and interaction with the solution occur on the surface, which may strongly affect the measured shell thickness. Furthermore, this value is in excellent agreement with previous results obtained by transmission electron microscopy (Figure 6). Special attention should be given to the width of the distribution curves under consideration. Dynamic light scattering curves have a long tail of up to ≈50 nm, which is absent in the data from the transmission electron microscopy. This discrepancy may be due to the fixing of the total hydrodynamic radius of small groups of the coated particles. However, another explanation is that a large sample of particles with dynamic light scattering and the registration of small amount of larger particles is taking place.

During the experiments, it was found that, under optimal conditions, the stirring speed does not affect the size of the obtained particles. However, the order of mixing is of great importance. Narrow particle size distribution and a small particle size is obtained only by the method of 8, which was described earlier.

Figure 8 shows the impact of the synthesis temperature on particle size distribution characteristics. From this it follows that the temperature of synthesis does not influence the numerical distribution of the particle sizes. However, it does have a very strong influence on the particle size distribution width and the intensity distribution. As the synthesis temperature rises, a second mode appears in the distribution corresponding to the particle sizes within the 80–100 nm range. This finding corresponds to the quantity of increase of large particles, but they still have an extremely small quantity compared to the total. This is directly associated with the effect of the sodium borohydride hydrolysis reaction and does not take into account changes in the $NaBH_4 : AgNO_3$ weight ratio. Consequently, the synthesis temperature is an extremely important condition that must be monitored to ensure that no impurities of large particles occur in the sols. It is worth mentioning that, after the stabilisation process, the resulting sol nanoparticles have a wide temperature range of stability. In these experiments, the particle size and optical properties are not changed in the temperature range from 0 to 60°C. The only limitations of the temperature application to the sol can be that a decrease in activity over long periods (>60 min) impacts temperatures > 60°C, at which the enzyme inactivation occurs [35]. Presumably, with partial destruction of the shell, there is a reduction of the stabilising ability. However, to our knowledge, up to a temperature of 90°C and an exposure time <60 min, changes in the parameters of the sol were not observed.

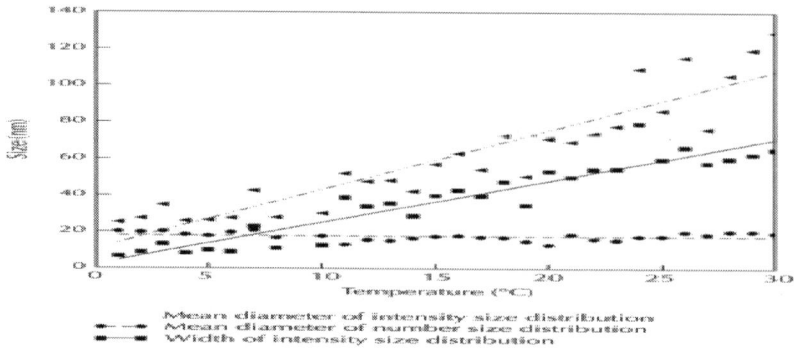

Figure 8: Dependence of the particle size distribution parameters on the synthesis temperature determined at the weight ratio $NaBH_4 : LYZ : AgNO_3 = 0.22 : 0.77 : 1$.

CONCLUSIONS

Our investigation has demonstrated the high importance of a comprehensive study of nanoparticle synthesis conditions.

The data obtained on $NaBH_4 : LYZ : AgNO_3$ aqueous solution systems show that the proper selection of the reagent ratio, synthesis temperature, and mixing speed makes it possible to obtain outstanding results on such qualitative parameters as the sol stability, mean particle size, and the size distribution width. In this study, we succeeded in establishing basic synthesis regularities that provide production of hybrid materials without additional purification and extraction of the necessary disperse particle fraction.

A problem of interest that calls for a subsequent, more profound study is the state and structure of a lysozyme molecule in the shell, and further research can be applied in the field of biosensors and drug development of a new generation.

ACKNOWLEDGMENTS

The authors gratefully acknowledge Andriyanova A. Yu for great help during the experimental work. Also thanks are due to Maslennikova

T. P. for assistance in the microscopic studies. Financial support was received from the Russian Foundation for Basic Research (Project no. 14-03-00626).

REFERENCES

1. M. Rai, A. Yadav, and A. Gade, "Silver nanoparticles as a new generation of antimicrobials,"Biotechnology Advances, vol. 27, no. 1, pp. 76–83, 2009.

2. N. Vigneshwaran, R. P. Nachane, R. H. Balasubramanya, and P. V. Varadarajan, "A novel one-pot "green" synthesis of stable silver nanoparticles using soluble starch," Carbohydrate Research, vol. 341, no. 12, pp. 2012–2018, 2006.

3. A. L. González and C. Noguez, "Optical properties of silver nanoparticles," Physica Status Solidi C, vol. 4, no. 11, pp. 4118–4126, 2007.

4. D. D. Evanoff Jr. and G. Chumanov, "Synthesis and optical properties of silver nanoparticles and arrays," ChemPhysChem, vol. 6, no. 7, pp. 1221–1231, 2005.

5. L. Balan, J.-P. Malval, R. Schneider, and D. Burget, "Silver nanoparticles: new synthesis, characterization and photophysical properties," Materials Chemistry and Physics, vol. 104, no. 2-3, pp. 417–421, 2007.

6. S. Shrivastava, T. Bera, S. K. Singh, G. Singh, P. Ramachandrarao, and D. Dash, "Characterization of antiplatelet properties of silver nanoparticles," ACS Nano, vol. 3, no. 6, pp. 1357–1364, 2009.

7. M. Liang, L. Wang, R. Su et al., "Synthesis of silver nanoparticles within cross-linked lysozyme crystals as recyclable catalysts for 4-nitrophenol reduction," Catalysis Science and Technology, vol. 3, no. 8, pp. 1910–1914, 2013.

8. O. Golubeva, O. Shamova, D. Orlov, E. Yamshchikova, A. Boldina, and V. Kokryakov, "Synthesis and investigation of silver-peptide bioconjugates and investigation in their antimicrobial activity," inMaterials Challenges and Testing for Supply of Energy and Resources, T. Böllinghaus, J. Lexow, T. Kishi, and M. Kitagawa, Eds., pp. 163–171, Springer, Berlin, Germany, 2012.

9. A. R. L. Caires, L. R. Costa, and J. Fernandes, "A close analysis of metal-enhanced fluorescence of tryptophan induced by silver nanoparticles: wavelength emission dependence," Central European Journal of Chemistry, vol. 11, no. 1, pp. 111–115, 2013.

10. I. Sondi and B. Salopek-Sondi, "Silver nanoparticles as antimicrobial agent: a case study on E. coli as a model for Gram-negative bacteria," Journal of Colloid and Interface Science, vol. 275, no. 1, pp. 177–182, 2004.

11. S. Chernousova and M. Epple, "Silver as antibacterial agent: ion, nanoparticle, and metal," Angewandte Chemie, vol. 52, no. 6, pp. 1636–1653, 2013.

12. S. R. Mudshinge, A. B. Deore, S. Patil, and C. M. Bhalgat, "Nanoparticles: emerging carriers for drug delivery," Saudi Pharmaceutical Journal, vol. 19, no. 3, pp. 129–141, 2011.

13. D. Joshi and R. K. Soni, "Laser-induced synthesis of silver nanoparticles and their conjugation with protein," Applied Physics A, vol. 116, no. 2, pp. 635–641, 2014.

14. M. L. Gulrajani, D. Gupta, S. Periyasamy, and S. G. Muthu, "Preparation and application of silver nanoparticles on silk for imparting antimicrobial properties," Journal of Applied Polymer Science, vol. 108, no. 1, pp. 614–623, 2008.

15. K. C. Song, S. M. Lee, T. S. Park, and B. S. Lee, "Preparation of colloidal silver nanoparticles by chemical reduction method," Korean Journal of Chemical Engineering, vol. 26, no. 1, pp. 153–155, 2009.

16. D. M. Eby, N. M. Schaeublin, K. E. Farrington, S. M. Hussain, and G. R. Johnson, "Lysozyme catalyzes the formation of antimicrobial silver nanoparticles," ACS Nano, vol. 3, no. 4, pp. 984–994, 2009.

17. T.-H. Kim, M. Kim, H.-S. Park, U. S. Shin, M.-S. Gong, and H.-W. Kim, "Size-dependent cellular toxicity of silver nanoparticles," Journal of Biomedical Materials Research Part A, vol. 100, no. 4, pp. 1033–1043, 2012.

18. C. Carlson, S. M. Hussein, A. M. Schrand et al., "Unique cellular interaction of silver nanoparticles: size-dependent generation of reactive oxygen species," Journal of Physical Chemistry B, vol. 112, no. 43, pp. 13608–13619, 2008.

19. S. Prabhu and E. Poulose, "Silver nanoparticles: mechanism of antimicrobial action, synthesis, medical applications, and toxicity effects," International Nano Letters, vol. 2, pp. 1–10, 2010.

20. S. F. Sweeney, G. H. Woehrle, and J. E. Hutchison, "Rapid purification and size separation of gold nanoparticles via diafiltration," Journal of the American Chemical Society, vol. 128, no. 10, pp. 3190–3197, 2006.

21. J. Benavides, O. Aguilar, B. H. Lapizco-Encinas, and M. Rito-Palomares, "Extraction and purification of bioproducts and nanoparticles using aqueous two-phase systems strategies," Chemical Engineering and Technology, vol. 31, no. 6, pp. 838–845, 2008.

22. V. S. Gurin, V. P. Petranovskii, and N. E. Bogdanchikova, "Metal clusters and nanoparticles assembled in zeolites: an example of stable materials with controllable particle size," Materials Science and Engineering C, vol. 19, no. 1-2, pp. 327–331, 2002.

23. A. Taleb, C. Petit, and M. P. Pileni, "Synthesis of highly monodisperce silver nanoparticles from aot reverse micelles: a way to 2d and 3d self organization," Chemistry of Materials, vol. 9, no. 4, pp. 950–959, 1997.

24. P. Mukherjee, A. Ahmad, D. Mandal, et al., "Fungus-mediated synthesis of silver nanoparticles and their immobilization in the mycelial matrix: a novel biological approach to nanoparticle synthesis," Nano Letters, vol. 1, no. 10, pp. 515–519, 2001.

25. R. Das, R. Jagannathan, C. Sharan, U. Kumar, and P. Poddar, "Mechanistic study of surface functionalization of enzyme lysozyme synthesized Ag and Au nanoparticles using surface enhanced Raman spectroscopy," Journal of Physical Chemistry C, vol. 113, no. 52, pp. 21493–21500, 2009.

26. G. K. Chandra, D. R. Tripathy, S. Dasgupta, and A. Roy, "Interaction of (−)-epigallocatechin gallate with lysozyme-conjugated silver nanoparticles," Applied Spectroscopy, vol. 66, no. 7, pp. 744–749, 2012.

27. T. Yang, Z. Li, L. Wang, C. Guo, and Y. Sun, "Synthesis, characterization, and self-assembly of protein lysozyme monolayer-stabilized gold nanoparticles," Langmuir, vol. 23, no. 21, pp. 10533–10538, 2007.

28. A. Panuszko, M. Wojciechowski, P. Bru dziak, P. W. Rakowska, and J. Stangret, "Characteristics of hydration water around hen egg lysozyme as the protein model in aqueous solution. FTIR spectroscopy and molecular dynamics simulation," Physical Chemistry Chemical Physics, vol. 14, no. 45, pp. 15765–15773, 2012.

29. E. Jones, T. Oliphant, P. Peterson, et al., "SciPy: Open source scientific tools for Python," 2001, /http://www.scipy.org.

30. V. Amendola, O. M. Bakr, and F. Stellacci, "A study of the surface plasmon resonance of silver nanoparticles by the discrete dipole approximation method: effect of shape, size, structure, and assembly," Plasmonics, vol. 5, no. 1, pp. 85–97, 2010.

31. G. Chandra, K. S. Ghosh, S. Dasgupta, and A. Roy, "Evidence of conformational changes in adsorbed lysozyme molecule on silver colloids," International Journal of Biological Macromolecules, vol. 47, no. 3, pp. 361–365, 2010.

32. J. Hu, R. S. Sheng, Z. S. Xu, and Y. Zeng, "Surface enhanced raman spectroscopy of lysozyme,"Spectrochimica Acta A, vol. 51, no. 6, pp. 1087–1096, 1995.

33. E. Podstawka, Y. Ozaki, and L. M. Proniewicz, "Adsorption of S-S containing proteins on a colloidal silver surface studied by surface-enhanced Raman spectroscopy," Applied Spectroscopy, vol. 58, no. 10, pp. 1147–1156, 2004.

34. G. D. Chumanov, R. G. Efremov, and I. R. Nabiev, "Surface-enhanced Raman spectroscopy of biomolecules. Part I.—water-soluble proteins, dipeptides and amino acids," Journal of Raman Spectroscopy, vol. 21, no. 1, pp. 43–48, 1990

35. B. Fischer, I. Sumner, and P. Goodenough, "Renaturation of lysozyme-temperature dependent of renaturation rate, renaturation yield, and aggregation: identification of hydrophobic folding intermediates," Archives of Biochemistry and Biophysics, vol. 306, no. 1, pp. 183–187, 1993.

Chapter 2

Surface-Enhanced Raman Scattering as an Emerging Characterization and Detection Technique

Mustafa Culha[1], Brian Cullum[2], Nickolay Lavrik[3], and Charles K. Klutse[2]

[1]Department of Genetics and Bioengineering, Yeditepe University, Atasehir, Kadıkoy, 34755 Istanbul, Turkey

[2]Department of Chemistry and Biochemistry, University of Maryland, Baltimore County, 1000 Hilltop Circle, Baltimore, MD 21250, USA

[3]Center for Nanophase Materials Sciences, Oak Ridge National Laboratory, Oak Ridge, TN 37831, USA

ABSTRACT

While surface-enhanced Raman spectroscopy (SERS) has been attracting a continuously increasing interest of scientific community since its discovery, it has enjoyed a particularly rapid growth in the last decade.

Most notable recent advances in SERS include novel technological approaches to SERS substrates and innovative applications of SERS in medicine and molecular biology. While a number of excellent reviews devoted to SERS appeared in the literature over the last two decades, we will focus this paper more specifically on several promising trends that have been highlighted less frequently. In particular, we will briefly overview strategies in designing and fabricating SERS substrates using deterministic patterning and then cover most recent biological applications of SERS.

INTRODUCTION

SERS is a Raman spectroscopic technique, which takes advantage of localized surface plasmon resonance (LSPR) in nanoscale systems based on coinage metals, such as silver and gold. As a result, Raman cross-sections of molecules on SERS-active substrates can reach values comparable to those of fluorescence spectroscopy [1–4]. Respectively, enhancements of Raman scattering signals ranging from a factor of 10^6 to 10^{16} have been reported in various studies [5–7]. The mechanisms for this enhancement in SERS are attributed largely to electromagnetic field enhancement due to LSPR and also to chemical interactions of the analyte with the substrate. As a result of these enhancements, SERS can have the sensitivity sufficient for the detection of ultratrace levels of analytes down to single molecules [3, 8]. This level of sensitivity is particularly useful for the detection of a small number of analyte molecules normally encountered in a single cell. Apart from the sensitivity, the inherent unique attributes of Raman are retained in SERS. Thus, it distinguishes vibrational signatures of molecular bonds, enabling label-free positive identification of analytes in complex cellular environments. Analyte detection can be multiplexed without spectral overlap because SERS provides spectra with narrow bandwidth. It is also very sensitive to slight changes in the orientation and structure of the molecules, allowing for structural elucidation. These characteristics, coupled with the weak Raman scattering of water, make SERS an ideal technique for analyzing complex biological samples that require little or no sample preparation. Importantly, SERS is achieved using a wide range of excitation frequencies, allowing for the selection of less energetic excitation (NIR to red) in order to reduce photodamage and background autofluorescence. Additionally, the analyte detection takes place at a close proximity to the SERS-active metal surface, thereby further reducing background autofluorescence through quenching. Therefore, SERS addresses most of the challenges of fluorescence detection for biological applications while providing comparable sensitivity.

Nanostructured metal surfaces or metal nanoparticle assemblies (commonly referred to as SERS substrates) are required for SERS measurements and their SERS activity and measurement reproducibility largely influence the extent of SERS applications. As a result of this, the development of high-performing SERS substrates is an ongoing investigation.

A remarkably wide variety of SERS active substrates and media have been explored in the last few decades. While roughened noble metal surfaces [9] and noble metal colloids were among historically first SERS active objects, more complex and optimized plasmonic systems prepared by using elaborate chemical synthesis [10–12], template-directed deposition [13–15] and nanosphere self-assembly [16, 17] have subsequently set the standard in the area of SERS active substrates. In general, all technological strategies of creating SERS active substrates can be broadly divided into the three main categories: (i) chemical synthesis, (ii) template-assisted methods, and (iii) deterministic (lithographic) patterning. The former two strategies can be referred to as a "bottom-up" approach, while lithographic patterning is a typical "top-down" approach. Since each of these strategies has its own advantages and limitations, a notable recent trend in creating SERS substrates with improved performance consists in combining top-down and bottom-up technological strategies [18].

Metal colloids in suspensions or aggregation have widely been used for SERS measurements due to their ease of preparation by chemical synthesis and the high SERS activities they exhibit especially at the interparticle spacing [19–22]. SERS enhancement factors as much as 10^{16} have been reported with such substrates, allowing for single-molecule detection [23–26]. When these colloids are immobilized on solid supports using chemical or electrostatic interaction, their reproducibility is tremendously improved. This is due to the ability to control interparticle spacing between the immobilized metal colloids. Additionally, such control of interparticle spacing has been exploited in the investigation of some fundamental concepts of SERS [27–29]. Template-assisted SERS substrates development have largely been aided by the advances in nanotechnology, leading to existence of a broad range of such substrates in recent years. One advantage of this group of substrates involves the use of nanostructure templates to control the interparticle distance and improve the reproducibility. Nanosphere lithography, for example, involves the deposition of a metal film on nanosphere templates arranged on a solid platform followed by the removal of the template. This leaves a regular of array of SERS-active metal nanostructures on the solid platforms [30, 31]. Substrates derived from metal film on nanostructures (MFON) are another template method except in this case; the nanostructures are not removed after the metal deposition. Thus, both the underlying nanostructures and the metal film deposited serve as the SERS substrates [32, 33]. These types of substrate are easily controlled based on the thickness of the metal film deposited and the size of the templates used, yielding highly reproducible SERS substrates. Importantly, by depositing several layers of SERS-active metal films separated by dielectric spacers, MFONs can be used in a multilayer geometry for improved SERS enhancement by as much as 2 orders of magnitude, showing that the SERS enhancement of conventional substrates can be further increased [34–36]. Since SERS discovery, a large variety of substrates has been reported for its measurements as captured in recent reviews [37–40], and all of them could not be covered in this paper. Although experimental work on SERS of biological objects has relied predominantly on "bottom up" approaches, such as colloidal assembly and synthesis, SERS substrates implemented using "top down" strategies, such as EBL, are likely to yield to the most reproducible SERS substrates in the future. Therefore, the following section will review several specific deterministic patterning strategies that offer great promise for fabricating high-performance SERS substrates.

SERS SUBSTRATES CREATED USING DETERMINISTIC PATTERNING STRATEGIES

It is worthy to note that, while chemical synthesis and template-directed methods have prevailed in creating intricate plasmonic structures with remarkable SERS performance [10, 41–44], deterministic patterning approaches, in particularly electron beam lithography (EBL), has been explored in the area of SERS far less extensively. This trend can be explained by the historically limited availability of EBL tools as well as the high cost of nanoscale lithographic processing compared to chemical synthesis. Nonetheless, notable examples of successful applications of EBL to SERS active substrates can be traced back to studies by Liao et al. in the early 1980s [45, 46]. A growing number of promising results on SERS active substrates prepared using lithography-derived processing have been reported more recently [18, 47–58].

In the section below, we will briefly discuss the recent advances in high-performance SERS substrates prepared using primarily EBL and wafer level cleanroom processing. Comprehensive discussions of conventional wafer level processes can be found in the literature [59]. Patterning of a substrate material by means of EBL is a multistep process that starts with designing patterns using computer-assisted design (CAD) software. EBL tool provides a programmable exposure of substrates coated with e-beam resist to a focused electron beam. As a result, the pattern is transferred onto a thin layer of e-beam resist that plays a role of a mask in the subsequent selective removal of the substrate material by dry or wet etching or in the subsequent selective removal of a deposited metal film by a liftoff process. Soft lithography relies on a patterned master to create morphologies complimentary to those present on the master in a single step, such as embossing or molding. Various modifications of anisotropic reactive ion etching (RIE) enable formation of grooves, wells, pillars, and other nonplanar structures on Si, SiO_2, and polymeric substrates with excellent control of the sidewall profile and characteristic sizes ranging from tens of nanometers [59]. As applied to SERS substrates, the main advantage of EBL patterning is its ability to create arbitrary 2D shapes with high fidelity on the spatial scale relevant to noble metal structures that exhibit localized plasmon resonance (LSPR) in the visible region of the electromagnetic spectrum [60]. These features, combined with the wider availability of EBL tools to the research community, have facilitated efforts towards combinatorial [60] as well as model-driven [18, 51, 61–63] SERS active substrates.

The most notable types of deterministically patterned SERS substrates include (i) dense periodic arrays and gratings, [35, 46, 61] (ii) plasmonic structures with extremely small (5 nm or less) nanogaps [48, 52, 64, 65], and (iii) multiscale structures and structures with complex 3D architectures [18, 51, 57, 58, 66]. Silver-coated gratings [61] and dense periodic arrays of nanoscale pillars [45, 46] were among historically first SERS substrates created by means of EBL. Fabrication of such substrates involved three main processing steps: EBL patterning of the resist layer, RIE of the substrate material to the desired depth (typically several hundred nanometers), and physical vapor deposition (PVD) of a 25 to 50 nm silver layer. The SERS enhancements of the resulting structures were reported to be at least one order of magnitude larger than that of silver island films [52]. Subsequently, several modifications of this processing sequence was adapted to create various highly optimized SERS substrates [55, 57, 58].

Plasmonic dimers shaped as nanoscale bowtie antennas [54] represent another interesting class of structures that can be precisely fabricated using EBL and utilized as very promising high-performance SERS substrates [52]. These substrates were characterized by SERS enhancement factor exceeding 10^{11}. An alternative approach to plasmonic structures with sub-10 nm gaps that give rise to very strong local enhancements of optical fields relied on photolithographic patterning and an alumina sacrificial spacer layer deposited using atomic layer deposition [55]. Such vertically oriented plasmonic nanogap arrays were characterized by local SERS enhancements of up to 10^9. Yet another type of SERS substrates with plasmonic dimers with few nm gaps was fabricated by creating discontinuities in EBL-patterned bridge structures using electromigration [48]. Electromigrated nanoscale gaps exhibited extremely strong SERS.

More recently, several modifications of the fabrication sequence that combines EBL, RIE, and PVD have been used to create sombrero-shaped [58] and disc-on-pillar plasmonic architectures further optimized to achieve strong local field enhancements [63].

SERS signal enhancements sufficient for detection of few molecules in vicinity of hot spots were reported for the sombrero-shaped structures [58] while the disk on-pillar structures exhibited averaged SERS enhancement factors above 10^9 [57].

A very interesting concept of aperiodic multiscale structures that takes advantage of the cascade enhancement effect was implemented using EBL patterning and subsequent in situ spatially selective reduction of gold. These two successive steps resulted in hierarchal formation of plasmonic nanoparticles with two characteristic sizes of approximately 200 nm and 30 nm, respectively. These multiscale structures exhibited reproducible spatially averaged SERS enhancements similar to 10^8 [18].

Although fabrication of the majority of the SERS active substrates discussed above relied on EBL, similar structures could be created using alternative approaches, such as optical interference [66, 67], nanostensil [68] lithographies, and laser-induced synthesis [68, 69]. Compared to these latter techniques, advantages of EBL include its wider availability and applicability to patterning both periodic [40] and aperiodic patterns [18, 50] of arbitrary shapes, a particularly useful feature at the research stage. A relatively low throughput and high cost of EBL tools makes them less attractive for scaled-up fabrication of SERS substrates. In order to make deterministic patterning of SERS substrates scalable, more flexible and less expensive advantages of EBL can be further augmented by nanoimprinting, nanoembossing [70], and nanotransfer stamping/printing approaches [53, 54]. For instance, a small-area silicon master with nanoscale patterns can be created using a combination of EBL and RIE and then used in a step-and-repeat process to pattern larger areas and/or a number of SERS substrates in a cost- and time-efficient manner.

DESIGN OF SERS-BASED NANOSENSORS FOR CELLULAR APPLICATIONS

In addition to lithographically produced or patterned substrate fabrication, SERS has also seen several significant advances in its application to biological and intracellular analyses over the pat decade. Such analyses take advantage of the inherent nanoscale size of the SERS active substrates to perform highly localized chemical-specific investigations. During the past couple of decades, such analyses have advanced from the pioneering work of inserting gold colloidal probes inside a cell for SERS-based chemical imaging of whatever species were present at the probes surface, to more recent versions of SERS-based sensors with various biological receptors for added specificity. In such analyses, the highly enhancing nature of the substrate must be retained as well as the addition of specific receptors. Figure 1 represents the general model of a typical SERS-based nanosensing showing a SERS nanosensor made of a SERS-active nanoparticle with a recognition elements tethered to it. The particle is a nanostructure small enough to be inserted into a living cell without significant perturbation or toxicity. It is also capable of supporting the induction of localized electromagnetic field when excited so that the molecular information of analyte closer to it can be obtained by SERS. Nanostructures made from gold and silver are often used as SERS-active particles because they provide the most enhanced SERS signals especially in the visible to NIR region. To promote analyte specificity, the SERS-active nanostructure can be attached with analyte recognition elements (e.g., enzymes, aptamers, and Fabs) or a Raman reporter. For intracellular analyte detection, the SERS nanosensor is inserted into the cell (through cellular uptake or physical injection). The targeted analyte interacts with the recognition elements and is brought closer to the SERS-active particle. The nanosensor is interrogated with an appropriate excitation source and the scattered radiation is collected and detected by a photodetector.

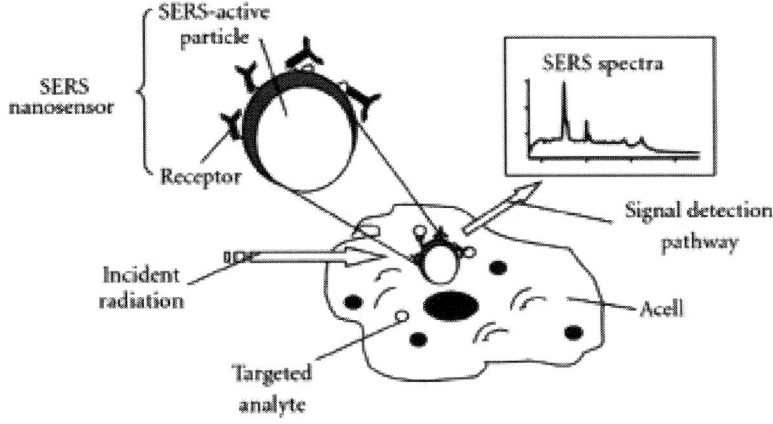

Figure 1: A general concept of SERS-based intracellular nanosensing.

Application of SERS in Intracellular Analyses

Colloidal plasmonic gold nanoparticles are commonly used for SERS-based intracellular analyses due to their small sizes, SERS-activity, and biocompatibility [71, 72]. Kneipp and coworkers were among the pioneers of SERS-based individual live cells monitoring [73–76]. In one study, gold nanoparticles with sizes ranging from 30 to 50 nm were introduced into immortalized rat renal proximal tubule (IRPT) cells and mouse microphage cell line (J774), via endocytosis. To minimize background autofluorescence, 786 nm was used for excitation source for the measurement of SERS spectra at different time points. Remarkably, it was noticed that the pattern of SERS fingerprints changed over time, with the spectra taken two hours after particles internalization showing the largest number of bands and highest intensity [63]. From these study, the potential of SERS as a sensitive method for label-free analyte detection inside an individual living cell was demonstrated. However, because any species in contact with the metal surface could potentially provide a signal, a significant background problem with such studies was found to exist. Furthermore, the strong interaction of some biological molecules with metal surfaces often resulted in biological fouling, as biomolecules randomly adsorbed onto the metal surfaces [73, 75, 77].

To selectively detect specific analytes in complex biological environments, nanoparticles have been immobilized with chemical receptors to form SERS-based nanosensors. Most of such SERS-based nanosensors have been developed for pH monitoring [78–82]. In one study, 4-mercaptobenzoic acid with pH sensitive COO⁻ vibration $(1430\,cm^{-1})$ was self-assembled on SERS-active nanoparticles. By monitoring the intensity of COO⁻ vibration, it was possible to monitor pH in Chinese hamster ovary cells at physiological concentrations [83]. Recently, functionalized gold nanoparticles have been used to monitor intracellular redox potential changes in NIH/3T3 fibroblast cells [75]. SERS-active nanoparticles made of 125 nm silica core and 25 nm thick gold shells were modified with redox-active SERS molecules (e.g. 2-mercaptobenzene-1, 2-diol). The redox-active molecules responded to the concentration of oxidative and reductive species in the cell through reversible redox reaction leading to slight changes in structure, which was reflected in SERS spectra [84].

SERS-Based Immunonanosensors

With increasing development of different types of SERS-active nanomaterials, it became clear that improvement in the specific analyte detection capability of these nanomaterials can rapidly advance SERS intracellular nanosensing technology. In view of this, various biorecognition molecules (including enzymes, Fabs, and aptamers) were earmarked for the development of SERS bionanosensors [85–88]. Efforts in this direction have led to the development of SERS bionanosensors employing antibodies for analyte recognition [89–92]. These SERS immunonanosensors developed by Cullum and coworkers employed metal film over nanostructure (MFON) SERS-active nanoparticles to provide uniform sensitivity [93] without the need for particle aggregation. Antibodies for specific antigens were immobilized on these nanoparticles to broaden the classes of analytes that can be interrogated to include proteins/peptides [91, 92] and other immunogenically recognized biomolecules (e.g., glucose, insulin, etc.) [89, 94].

Insulin receptive immunonanosensors were fabricated by immobilizing anti-human insulin receptors on MFON SERS-active particles via crosslinkers. Suitable crosslinkers were identified by characterizing several crosslinkers. Figure 2 shows the spectra of

2-mercapto-4-methyl-5-thiazoacetic acid (MMT), which was the preferred crosslinker because it exhibited rigidity and minimal spectral peaks while providing significant bands for use as internal standards. After attachment of the receptor (i.e., antibody) via traditional EDC chemistry, evaluation of the fabricated immunonanosensors in the cell culture media revealed significant and reproducible changes in the SERS spectra physiological concentrations of human insulin.

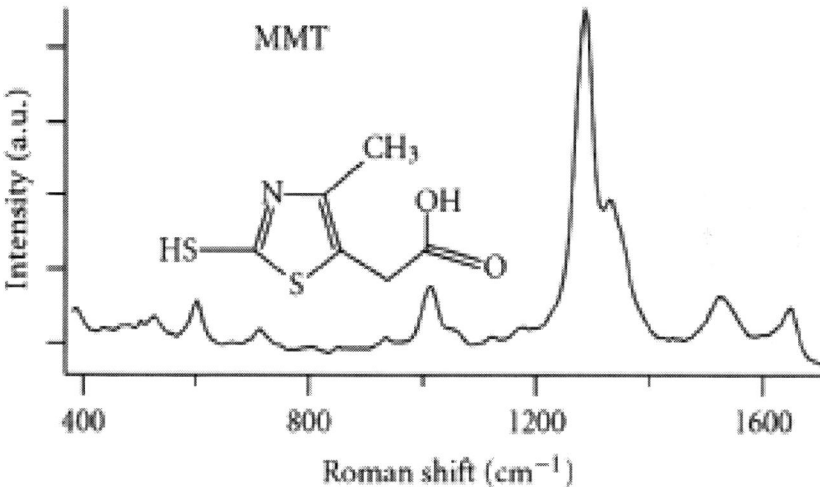

Figure 2: SERS spectra and structure of the cross-linker MMT.

This was because antibody-analyte interaction led to changes in the conformation of the antibody. The spectra showing the various stages of nanosensor evaluation are shown in Figure 3. That specific Raman bands (443, 808, 1625 cm⁻¹), which are associated with the antibody, were not present when the analyte interacted with the receptor. By attaching interleukin II (IL-2) receptors to SERS nanosensor surfaces through the same attachment process, it was possible to fabricate different types of nanosensors for the detection of IL-2 to a level as low as 10 μg/mL [92, 95] indicating that the nanosensors could be modified for a wide range of applications.

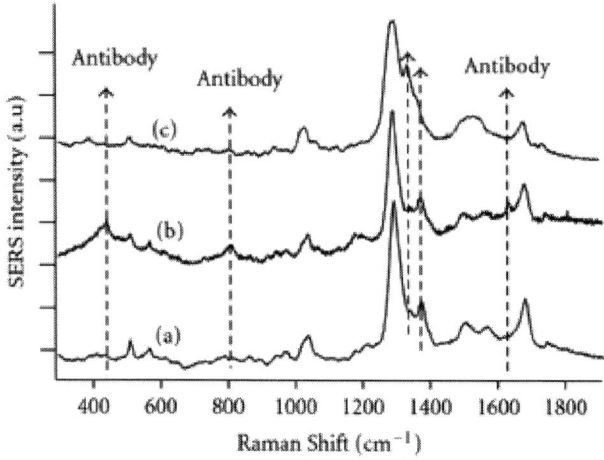

Figure 3: SERS spectra of (a) activated MMT bound to silver-coated nanosphere with 40% surface coverage, (b) after binding anti-human insulin to the MMT, and (c) of (b) (immunonanosensors) in the presence of 10 µg/mL of insulin.

SERS-Based Nanoimaging

Besides achieving sensitivity for ultratrace intracellular analyte detection, extracellular SERS imaging of cellular and bacterial surfaces has recently seen significant interest. Such extracellular imaging is capable of providing important chemical specific spatial information (e.g., size and distribution) of the macromolecules present. Such analyses significantly improve our understanding of how macromolecules are transported and assembled at specific sites particularly if rapid dynamic imaging is possible. To achieve such analyses, SERS methods with high-spatial resolution are required. A commonly used configuration for improving the spatial resolution of SERS is the tip-enhanced Raman spectroscopy (TERS). This takes advantage of a precise positioning of a nanotip SERS-active substrate to achieve localized SERS enhancement with subdiffraction-limited spatial resolution. TERS is normally achieved by modifying the tips of scanning probe microscopy (SPM) with SERS-active metals. For example, SERS has been coupled with near-field scanning optical microscopy (NSOM), to provide highly resolved chemical images of multiple trace species [96–98]. Other

SPMs that have been coupled with SERS to provide the subdiffraction-limited chemical imaging are atomic force microscopy AFM and scanning tunneling microscopy STM [99, 100]. For example, silver coated AFM-tip has been used to gather both topographical and SERS spectral information of a bacterial membrane. The SERS spectral showed the peptide and sugar components of the cell surface [100]. However, TERS is only useful for static images as it typically requires long periods of time to scan across the entire sampling area.

To provide a nanosensor/nanoprobe capable of obtaining dynamic images, a novel nanoimaging probe has been developed whereby coherent fiber optic imaging bundles were coupled with SERS for real-time subdiffraction-limited chemical imaging [68, 101–104]. This SERS nanoimaging probe has an advantage of providing both temporal and spatial resolution, thereby allowing the monitoring of the movement of macromolecules across membranes. The technique takes advantage of a tapered fiber optic with a large collection of fiber elements (30 000). The coherent nature of the packed bundle is retained during the tapering process, which is controlled in order to vary the diameter of the entire tip. This variation in diameter yields predictable spatial resolution that can be varied from microscopic to subdiffraction-limited levels. The SERS surface is produced on the tip through a controlled chemical etching process followed by selective deposition of silver or gold, which creates reproducible hexagonal spikes for the roughness (Figure 4). For this reason, uniform SERS enhancements across the surface of the probe are achieved.

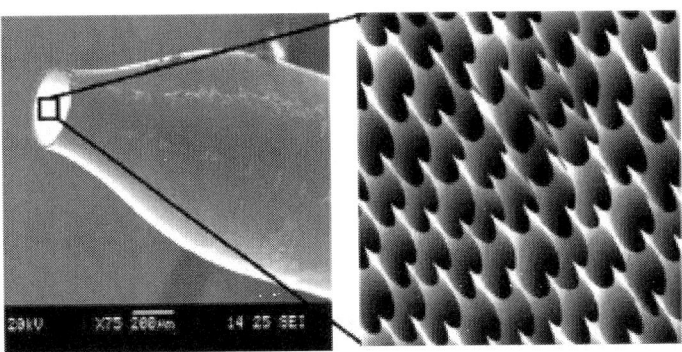

Figure 4: SEM images of (a) a tapered optic fiber probe and (b) its etched tip which is coated with SERS-active metal.

In proof-of-principle studies, such probes have been used for chemical differentiation of various samples with <100 nm spatial resolution [68, 101–104]. In one such study, gelatin was homogenously mixed with brilliant cresyl blue (BCB), while benzoic acid was spotted at various locations around the edge. By tuning to the appropriate wavelength, each of the two analytes was selectively imaged in the sample as shown in the images and the spectra in Figure 5. Additionally, time-lapse images actually allowed for the diffusion of the chemicals to be monitored with millisecond temporal resolution [68, 101]. Using these SERS nanoimaging probes, dynamic visualization of biochemical species associated with cell membrane components have been measured, demonstrating the potential of such probes for providing a deeper understanding of the various physiological processes in cellular membranes [105].

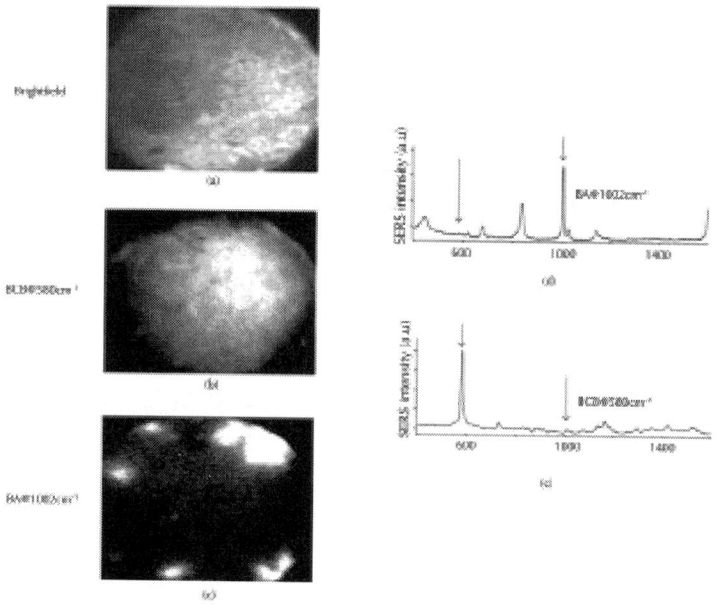

Figure 5: Chemical differentiation using SERS imaging probe tapered down to 140 nm. (a) A brightfield image of the gelatin, (b) a SERS image of the localized benzoic acid, and (c) a SERS image of the BCB, and (d) and (e) are the corresponding spectra of images (b) and (c), respectively.

DETECTION, IDENTIFICATION, AND CHARACTERIZATION OF BIOMACROMOLECULES AND MICROORGANISMS

There is an enormous interest to use SERS for the solution of problems in biorelated fields such as medicine, molecular biology, and microbiology. The advantage of SERS over other techniques such as fluorescence, IR, and mass spectroscopy is already stated above. Although all these techniques have distinct advantages, SERS could be a proper replacement for certain applications. For example, narrow-spectral bandwidth, detection and identification of a molecule without an external label, insensitivity to water, nondestructive nature, and low cost for instrumentation can be given the advantages over the mentioned techniques. However, the several issues inherent to the nature of the technique, mainly reproducibility, still persist and lower the applicability to its wide spread use in biological applications.

The detection and identification of several biological molecules such as proteins [106–110], DNA [111–114], and RNA [115, 116] and molecular organizations such as bacteria [117–126], yeast [127–129], and viruses [130, 131] using SERS were reported. The detection scheme could be based on either using the intrinsic fingerprint information of the molecules or molecular structures or the use of an external label. Although the use of the molecules' fingerprint spectrum is highly desirable for identification and detection, obtaining a healthy spectrum from a biological molecule such as proteins could be difficult. In this section, the detection and identification of proteins, DNA/RNA and whole microorganisms will be discussed. There are several reviews [132–134] in literature summarizing the developments in previous years and only most recent and some of the important reports are discussed here.

In most of the indirect protein detection assays, SERS is used as a replacement of either fluorescence or staining step. There are several reports regarding the use of the technique in immunoassays format for protein or microorganism detection [135–141]. In immunoassay-based approaches, a recognition element, an antibody, captures the target and a transducer converts the biological recognition to readout. The transducer could be a radiological, electrochemical, or optical. SERS is generally used as a replacement of these transducers types. The high sensitivity and the multiplexing properties of the technique can be given as its advantages over other transducer types.

In a one-step homogenous immunoassay, label-free detection of human IgG down to 0.1 μg/mL without using an external label was reported [126]. In another report, a photobleaching-resistant immunoassay system was developed for the detection of protein A, which is the surface antigen of Staphylococcus aureus. The achieved detection limit for this protein was 1 pg/mL [136].

In a recent study, a SERS-based sandwich immunoassay coupled with an optoelectronic microfluidic system for the detection of human tumor marker, alphafetoprotein, was reported. A detection limit down to 0.1 ng/mL using a 500 nL sample droplet in 5 minutes was accomplished [137]. Figure 6 illustrates the optoelectronic sandwich immunoassay procedure.

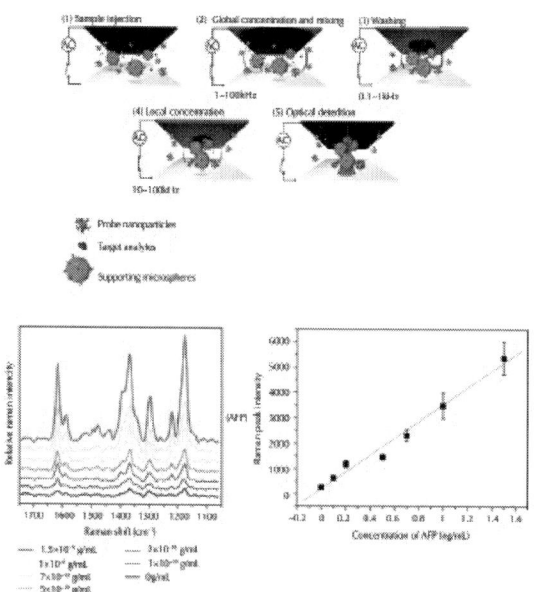

Figure 6: Illustration of optoelectronic sandwich immunoassays procedure, SERS spectra at decreasing AFP concentrations, and calibration curve at 1615 cm⁻¹, reprinted with permission from [137].

In another recent report, immunoassay-based SERS method for the detection of cancer marker, angiogenin (ANG), and alpha-fetoprotein (AFP), was successfully demonstrated with a detection limit of 0.1 pg/mL and 1.0 pg/mL, respectively [138]. Wang et al. demonstrated the detection of MUC4 in real samples, which is a pancreatic cancer marker, using a SERS-based immunoassay platform [139]. The detection of bacteria based on SERS immunoassay is also attracting interest. In a heterogeneous immunoassay, the detection of E. coli concentration in range of 10^1–10^5 cfu/mL was demonstrated [140]. The same group also demonstrated the detection of E coli down to 8 cfu mL^{-1} using a sandwich immunoassay [141].

The detection and identification of proteins have a critical importance in several fields including medicine, biotechnology, and pharmacology. The conventional approaches such as immunoassay-based techniques and mass spectroscopy are powerful but they have certain drawbacks such a low sensitivity with immunoassay-based techniques and high-cost with mass spectroscopic approach. With the proven high sensitivity, SERS can be a complementary technique, even alternative to some applications, for the protein detection and identification. As mentioned earlier, SERS can be used as replacement for fluorescence or radiolabeling. However, the potential of the technique for label-free detection and identification of bio macromolecules has not been fully explored yet. Among the biomcaromolecules, the proteins are the most important group, and their detection and identification have a great importance. Therefore, the use of SERS as label-free detection and identification of proteins is pursued. Han et al. recently reviewed the detection of proteins using SERS [142]. The biggest obstacle with the label-free detection of proteins is the diversity of the protein shape, size, and surface charge properties. When colloidal noble metal nanoparticles are used in the SERS experiment, it is very difficult to control the aggregation behavior of the protein-NP structures during drying process. When a simple mixing and drying approach is used for the sample preparation, only proteins with a chromophore/heme group provide reasonably rich spectra Otherwise, a very poor spectrum is obtained. Therefore, a more systematic approach must be followed to obtain good spectra from proteins. For example, Zhou et al. reported fast label-free semiquantitative detection of proteins down to submonolayer coverage using nitrate ion [143].

Kahraman et al. recently used "convective-assembly" approach for label-free detection of proteins regardless of their size, shape, and surface charge [108]. In this approach, mixture of protein-AgNP colloid is spotted at the cross-section of two glass slides, which one is placed on a moving stage and other is located with an angle of 23° on top of the other at a fixed position. Figure 7 shows the "convective-assembly" set-up and process. As the stage moves, the droplet at the cross-section is slowly dragged. During the dragging process, the evaporation forces the protein-AgNP structures to form a thin film in a more controlled manner. During the drying process AgNP and proteins are forced to stay close to each other, which has a dramatic effect on SERS spectra. Using this approach, a more reproducible sample preparation was possible. The detection limit using this approach was estimated as 0.50 µg/mL for all proteins used in the study, which was about one order of magnitude lower than the previously reported detection limits. The same group later also demonstrated the differential separation of the binary and ternary protein mixtures using convective assembly process and detection and identification of proteins in these mixtures [110].

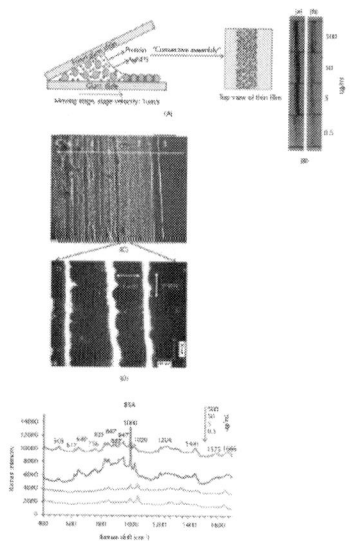

Figure 7: Illustration of process of convective assembly of protein-AgNP mixtures (A), photograph of lysozyme (a) and BSA (b) thin films (B), thin film

images of BSA-AgNP film structures under 5x (C) and 50x objectives (D). SERS spectra of BSA-AgNP with decreasing concentrations, reprinted with permission from [108].

Whole microorganism detection and identification using SERS is another research area where the fingerprint spectra obtained from whole microorganisms can be used. There are a number of studies demonstrating the feasibility of technique for whole bacterial detection and identification [117–126]. The identification and classification of bacteria causing urinary tract infections were demonstrated by Goodacre group using citrate reduced AgNPs [123]. The use of gold-coated silica nanoparticles as substrate and a barcoding approach was later used for the bacterial identification by Zeigler group [144]. The major problem with the application of SERS as a technique for microorganism identification is the spectral reproducibility. When the fact that the microorganisms are living systems is taken into account besides the number of parameters influencing the ultimate spectrum due to the nature of the technique, it becomes more confusing. Therefore, a solid protocol is obligatory for healthy interpretation of the obtained results. In this regard, Kahraman et al. followed several different approaches to obtained reproducible spectra from bacteria and studied the parameters influencing spectral reproducibility such as laser wavelength and colloidal nanoparticle concentration used for sample preparation [145].

For example increasing the colloidal AgNP concentration before mixing with the bacterial sample improved the sample-to-sample spectral reproducibility [126]. In another example, the bacteria and AgNPs were assembled into a thin film with "convective-assembly" method in order to generate a more uniform sample on a surface [146]. Figure 8 shows the SEM image of a sample assembled onto a glass surface using convective assembly. As it was stated above, there are several parameters that may affect the SERS spectra of microorganisms. In addition to the parameters pertaining to the SERS experimental conditions such as substrate and laser wavelength, the microorganisms are living things and they may show variations in their biochemical structure as they continue to grow in their life cycle. Besides the reproducibility issues, the origin of the spectral bands has not been completely understood yet.

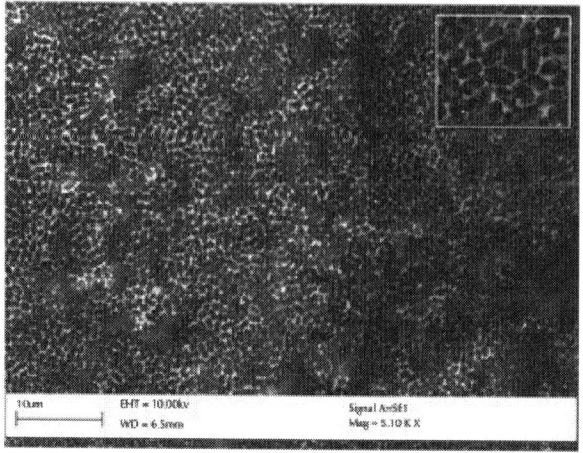

Figure 8: "Convectively assembled" E. coli on glass surface, reprinted with permission from [146].

When colloidal AgNPs or AuNPs are mixed with a complex biological sample, the molecular or ionic species that have an affinity for the noble metal surface preferentially interacts with the metal surface. From this point, one can easily conclude that the SERS spectra from such a complex mixture are dominated by the vibrational bands originating from those molecules or species preferentially bound to the metal surface. This leads to the striking similarity in the spectral pattern of biological samples, even though they are completely from different origins. For example, the growth media for bacterial has striking similarity with some bacteria such as E. coli. Keeping this point in mind, one needs to be very careful with the experimental design and should have a very good knowledge of the sample composition and sample handling for a healthy interpretation of the experimental result. The origin of the bands that appeared on the SERS spectra of bacteria was independently investigated by two groups. Kahraman et al. demonstrated that the SERS spectra obtained from a bacterial sample after several washing steps originate from the bacterial structures on the bacteria wall with some contribution from metabolites released during the sample preparation, mixing, and drying [147]. Premasiri et al. showed that the bacteria grown in different culture media resulted with the same SERS spectrum indicating the source of spectral features on the spectrum were from bacteria [148].

CONCLUSIONS

The investigation and optimization of high performing SERS substrates remains a very active area pursued by a number of cross-disciplinary research teams. The evidence, that SERS can be used for single-molecule detection and clear understanding of the relationship between LSPR and SERS enhancement mechanism, has inspired many recent studies focused on design and fabrication of new types of SERS substrates [149–151].

A number of recent studies indicate that SERS is a viable analytical technique for detection and differentiation of chemical species in biological cells and biologically relevant samples. Chemically synthesized noble metal nanoparticles and colloidal systems have been among SERS substrates with the highest reported values of SERS signal enhancement. Chemically synthesized noble metal colloids are also the most suitable SERS substrates for biological assays. This is largely due to their stability in aqueous media, inexpensive synthesis, and wide availability of precursor chemicals. On the other hand, a combination of deterministic (lithographic) patterning and wafer level processing offers a very attractive alternative route to SERS substrates that can be rationally designed and further optimized using theoretical models. It can be anticipated that this latter approach will ultimately lead to highly reproducible SERS substrates suitable for analysis of biological samples.

Although applications of lithographically patterned SERS substrates to detection of biological analytes remain relatively rare, ongoing efforts that bridge distinct technological strategies will likely bring a series of breakthroughs in this area in the nearest future. Already an impressive reproducibility and stability have been achieved in SERS analysis of biological samples using recently implemented SERS substrates. In many cases, however, the sensitivity needs to be improved in order to achieve the same level of success as with fractal colloidal aggregates. Further substantial advances can also be anticipated in SERS-based imaging with applications to live biological cells and microorganisms.

ACKNOWLEDGMENTS

N. Lavrik acknowledges support by the Scientific User Facilities Division, Office of Basic Energy Sciences, U.S. Department of Energy. M. Culha acknowledges support by Scientific and Technological Research Council of Turkey (TUBITAK) and Yeditepe University.

REFERENCES

1. A. Campion and P. Kambhampati, "Surface-enhanced Raman scattering," Chemical Society Reviews, vol. 27, no. 4, pp. 241–250, 1998.

2. A. Otto, "What is observed in single molecule SERS, and why?" Journal of Raman Spectroscopy, vol. 33, no. 8, pp. 593–598, 2002. · ·

3. K. Kneipp, Y. Wang, H. Kneipp et al., "Single molecule detection using surface-enhanced Raman scattering (SERS)," Physical Review Letters, vol. 78, no. 9, pp. 1667–1670, 1997.

4. I. Pavel, E. McCarney, A. Elkhaled, A. Morrill, K. Plaxco, and M. Moskovits, "Label-free SERS detection of small proteins modified to act as bifunctional linkers," Journal of Physical Chemistry C, vol. 112, no. 13, pp. 4880–4883, 2008. · ·

5. K. L. Kelly, E. Coronado, L. L. Zhao, and G. C. Schatz, "The optical properties of metal nanoparticles: the influence of size, shape, and dielectric environment," Journal of Physical Chemistry B, vol. 107, no. 3, pp. 668–677, 2003. · ·

6. C. L. Haynes and R. P. Van Duyne, "Plasmon-sampled surface-enhanced Raman excitation spectroscopy," Journal of Physical Chemistry B, vol. 107, no. 30, pp. 7426–7433, 2003.

7. S. Nie and S. R. Emory, "Probing single molecules and single nanoparticles by surface-enhanced Raman scattering," Science, vol. 275, no. 5303, pp. 1102–1106, 1997. · ·

8. M. E. Hankus, H. Li, G. J. Gibson, and B. M. Cullum, "Surface-enhanced Raman scattering-based nanoprobe for high-resolution, non-scanning chemical imaging," Analytical Chemistry, vol. 78, no. 21, pp. 7535–7546, 2006. · ·

9. M. Moskovits, "Surface roughness and the enhanced intensity of Raman scattering by molecules adsorbed on metals," The Journal of Chemical Physics, vol. 69, no. 9, pp. 4159–4161, 1978.

10. D. K. Lim, K. S. Jeon, H. M. Kim, J. M. Nam, and Y. D. Suh, "Nanogap-engineerable raman-active nanodumbbells for single-molecule detection," Nature Materials, vol. 9, no. 1, pp. 60–67, 2010. · ·

11. L. J. Sherry, S. H. Chang, G. C. Schatz, R. P. Van Duyne, B. J. Wiley, and Y. Xia, "Localized surface plasmon resonance spectroscopy of single silver nanocubes," Nano Letters, vol. 5, no. 10, pp. 2034–2038, 2005. · ·

12. L. J. Sherry, R. Jin, C. A. Mirkin, G. C. Schatz, and R. P. Van Duyne, "Localized surface plasmon resonance spectroscopy of single silver triangular nanoprisms," Nano Letters, vol. 6, no. 9, pp. 2060–2065, 2006. · ·

13. S. J. Hurst, E. K. Payne, L. Qin, and C. A. Mirkin, "Multisegmented one-dimensional nanorods prepared by hard-template synthetic methods," Angewandte Chemie, vol. 45, no. 17, pp. 2672–2692, 2006. · ·

14. L. D. Qin, S. Park, L. Huang, and C. A. Mirkin, "Matrials science: on-wire lithography," Science, vol. 309, no. 5731, pp. 113–115, 2005. · ·

15. H. Ko and V. V. Tsukruk, "Nanoparticle-decorated nanocanals for surface-enhanced Raman scattering,"Small, vol. 4, no. 11, pp. 1980–1984, 2008. · ·

16. C. L. Haynes and R. P. Van Duyne, "Plasmon-sampled surface-enhanced Raman excitation spectroscopy," Journal of Physical Chemistry B, vol. 107, no. 30, pp. 7426–7433, 2003.

17. C. L. Haynes and R. P. Van Duyne, "Nanosphere lithography: a versatile nanofabrication tool for studies of size-dependent nanoparticle optics," Journal of Physical Chemistry B, vol. 105, no. 24, pp. 5599–5611, 2001. · ·

18. A. Gopinath, S. V. Boriskina, W. R. Premasiri, L. Ziegler, B. M. Reinhard, and L. D. Negro, "Plasmonic nanogalaxies: multiscale aperiodic arrays for surface-enhanced Raman sensing," Nano Letters, vol. 9, no. 11, pp. 3922–3929, 2009. · ·

19. K. Kneipp, H. Kneipp, I. Itzkan, R. R. Dasari, and M. S. Feld, "Ultrasensitive chemical analysis by Raman spectroscopy," Chemical Reviews, vol. 99, no. 10, pp. 2957–2975, 1999.

20. K. Kneipp, A. S. Haka, H. Kneipp et al., "Surface-enhanced raman spectroscopy in single living cells using gold nanoparticles," Applied Spectroscopy, vol. 56, no. 2, pp. 150–154, 2002. · ·

21. J. B. Jackson and N. J. Halas, "Surface-enhanced Raman scattering on tunable plasmonic nanoparticle substrates," Proceedings of the National Academy of Sciences of the United States of America, vol. 101, no. 52, pp. 17930–17935, 2004. · ·

22. M. Rycenga, P. H. C. Camargo, W. Li, C. H. Moran, and Y. Xia, "Understanding the SERS effects of single silver nanoparticles and their dimers, one at a time," Journal of Physical Chemistry Letters, vol. 1, no. 4, pp. 696–703, 2010. · ·

23. K. Kneipp, Y. Wang, H. Kneipp et al., "Single molecule detection using surface-enhanced Raman scattering (SERS)," Physical Review Letters, vol. 78, no. 9, pp. 1667–1670, 1997.

24. S. M. Nie and S. R. Emory, "Probing single molecules and single nanoparticles by surface-enhanced Raman scattering," Science, vol. 275, no. 5303, pp. 1102–1106, 1997. · ·

25. S. Corni and J. Tomasi, "Surface enhanced Raman scattering from a single molecule adsorbed on a metal particle aggregate: a theoretical study," Journal of Chemical Physics, vol. 116, no. 3, pp. 1156–1164, 2002. · ·

26. X. M. Qian and S. M. Nie, "Single-molecule and single-nanoparticle SERS: from fundamental mechanisms to biomedical applications," Chemical Society Reviews, vol. 37, no. 5, pp. 912–920, 2008. · ·

27. M. Baia, F. Toderas, L. Baia, D. Maniu, and S. Astilean, "Multilayer structures of self-assembled gold nanoparticles as a unique SERS and SEIRA substrate," ChemPhysChem, vol. 10, no. 7, pp. 1106–1111, 2009. · ·

28. J. F. Li, Y. F. Huang, Y. Ding et al., "Shell-isolated nanoparticle-enhanced Raman spectroscopy," Nature, vol. 464, no. 7287, pp. 392–395, 2010. · ·

29. Y. Han, R. Lupitskyy, T. M. Chou, C. M. Stafford, H. Du, and S. Sukhishvili, "Effect of oxidation on surface-enhanced

raman scattering activity of silver nanoparticles: a quantitative correlation," Analytical Chemistry, vol. 83, no. 15, pp. 5873–5880, 2011. · ·

30. J. C. Hulteen, D. A. Treichel, M. T. Smith, M. L. Duval, T. R. Jensen, and R. P. Van Duyne, "Nanosphere lithography: size-tunable silver nanoparticle and surface cluster arrays," Journal of Physical Chemistry B, vol. 103, no. 19, pp. 3854–3863, 1999. · ·

31. M. D. Malinsky, K. Lance Kelly, G. C. Schatz, and R. P. van Duyne, "Nanosphere lithography: effect of substrate on the localized surface plasmon resonance spectrum of silver nanoparticles," Journal of Physical Chemistry B, vol. 105, no. 12, pp. 2343–2350, 2001.

32. L. A. Dick, A. D. McFarland, C. L. Haynes, and R. P. Van Duyne, "Metal film over nanosphere (MFON) electrodes for surface-enhanced Raman spectroscopy (SERS): improvements in surface nanostructure stability and suppression of irreversible loss," Journal of Physical Chemistry B, vol. 106, no. 4, pp. 853–860, 2002. · ·

33. J. Stropp, G. Trachta, G. Brehm, and S. Schneider, "A new version of AgFON substrates for high-throughput analytical SERS applications," Journal of Raman Spectroscopy, vol. 34, no. 1, pp. 26–32, 2003. · ·

34. H. Li and B. M. Cullum, "Dual layer and multilayer enhancements from silver film over nanostructured surface-enhanced Raman substratess," Applied Spectroscopy, vol. 59, no. 4, pp. 410–417, 2005.

35. H. Li, C. E. Baum, J. Sun, and B. M. Cullum, "Multilayer enhanced gold film over nanostructure surface-enhanced Raman," Applied Spectroscopy, vol. 60, no. 12, pp. 1377–1385, 2006. · ·

36. C. K. Klutse and B. M. Cullum, "Optimization of SAM-based multilayer SERS substrates for intracellular analyses: the effect of terminating functional groups," in Smart Biomedical and Physiological Sensor Technology VIII, vol. 8025 of Proceedings of SPIE 8025, Orlando, Fla, USA, April 2011.

37. B. M. Cullum, H. Li, M. V. Schiza, and M. E. Hankus, "Characterization of multilayer-enhanced surface-enhanced Raman scattering (SERS) substrates and their potential for SERS nanoimaging,"Nanobiotechnology, vol. 3, no. 1, pp. 1–11, 2007.

38. J. A. Dieringer, A. D. McFarland, N. C. Shah et al., "Surface enhanced Raman spectroscopy: new materials, concepts, characterization tools, and applications," Faraday Discussions, vol. 132, pp. 9–26, 2006. · ·

39. K. Hering, D. Cialla, K. Ackermann et al., "SERS: a versatile tool in chemical and biochemical diagnostics," Analytical and Bioanalytical Chemistry, vol. 390, no. 1, pp. 113–124, 2008. · ·

40. M. Fan, G. F. S. Andrade, and A. G. Brolo, "A review on the fabrication of substrates for surface enhanced Raman spectroscopy and their applications in analytical chemistry," Analytica Chimica Acta, vol. 693, no. 1-2, pp. 7–25, 2011.

41. L. Qin, S. Zou, C. Xue, A. Atkinson, G. C. Schatz, and C. A. Mirkin, "Designing, fabricating, and imaging Raman hot spots," Proceedings of the National Academy of Sciences of the United States of America, vol. 103, no. 36, pp. 13300–13303, 2006. · ·

42. W. Y. Li, P. H. C. Camargo, X. Lu, and Y. Xia, "Dimers of silver nanospheres: facile synthesis and their use as hot spots for surface-enhanced raman scattering," Nano Letters, vol. 9, no. 1, pp. 485–490, 2009. · ·

43. M. L. Pedano, S. Li, G. C. Schatz, and C. A. Mirkin, "Periodic electric field enhancement along gold rods with nanogaps," Angewandte Chemie, vol. 49, no. 1, pp. 78–82, 2010. · ·

44. R. Kodiyath, J. Wang, Z. A. Combs et al., "SERS effects in silver-decorated cylindrical nanopores," Small, vol. 7, no. 24, pp. 3452–3457, 2011. · ·

45. A. Wokaun, J. P. Gordon, and P. F. Liao, "Radiation damping in surface-enhanced Raman scattering," Physical Review Letters, vol. 48, no. 14, pp. 957–960, 1982. ·

46. P. F. Liao, J. G. Bergman, D. S. Chemla et al., "Surface-enhanced raman scattering from microlithographic silver particle surfaces," Chemical Physics Letters, vol. 82, no. 2, pp. 355–359, 1981.

47. M. A. De Jesús, K. S. Giesfeldt, J. M. Oran, N. A. Abu-Hatab, N. V. Lavrik, and M. J. Sepaniak, "Nanofabrication of densely packed metal-polymer arrays for surface-enhanced raman spectrometry," Applied Spectroscopy, vol. 59, no. 12, pp. 1501–1508, 2005. · ·

48. D. R. Ward, N. K. Grady, C. S. Levin et al., "Electromigrated nanoscale gaps for surface-enhanced Raman spectroscopy," Nano Letters, vol. 7, no. 5, pp. 1396–1400, 2007. · ·

49. N. A. Abu Hatab, J. M. Oran, and M. J. Sepaniak, "Surface-enhanced Raman spectroscopy substrates created via electron beam lithography and nanotransfer printing," ACS Nano, vol. 2, no. 2, pp. 377–385, 2008. · ·

50. S. M. Wells, S. D. Retterer, J. M. Oran, and M. J. Sepaniak, "Controllable nanofabrication of aggregate-like nanoparticle substrates and evaluation for surface-enhanced Raman spectroscopy," ACS Nano, vol. 3, no. 12, pp. 3845–3853, 2009. · ·

51. B. Yan, A. Thubagere, W. R. Premasiri, L. D. Ziegler, L. D. Negro, and B. M. Reinhard, "Engineered SERS substrates with multiscale signal enhancement: nanoparticle cluster arrays," ACS Nano, vol. 3, no. 5, pp. 1190–1202, 2009. · ·

52. N. A. Hatab, C. H. Hsueh, A. L. Gaddis et al., "Free-standing optical gold bowtie nanoantenna with variable gap size for enhanced Raman spectroscopy," Nano Letters, vol. 10, no. 12, pp. 4952–4955, 2010. · ·

53. D. Bhandari, I. I. Kravchenko, N. V. Lavrik, and M. J. Sepaniak, "Nanotransfer printing using plasma etched silicon stamps and mediated by in situ deposited fluoropolymer," Journal of the American Chemical Society, vol. 133, no. 20, pp. 7722–7724, 2011. · ·

54. D. Bhandari, S. M. Wells, A. Polemi, I. I. Kravchenko, K. L. Shuford, and M. J. Sepaniak, "Stamping plasmonic nanoarrays on SERS-supporting platforms," Journal of Raman Spectroscopy, vol. 42, no. 11, pp. 1916–1924, 2011. · ·

55. J. D. Caldwell, O. Glembocki, F. J. Bezares et al., "Plasmonic nanopillar arrays for large-area, high-enhancement surface-enhanced Raman scattering sensors," ACS Nano, vol. 5, no. 5, pp. 4046–4055, 2011. · ·

56. A. S. P. Chang, M. Bora, H. T. Nguyen et al., "Nanopillars array for surface enhanced Raman scattering," in Advanced Environmental, Chemical, and Biological Sensing Technologies VIII, vol. 8024 of Proceedings of the SPIE, pp. 802401–802409, 2011.

57. S. M. Wells, A. Polemi, N. V. Lavrik, K. L. Shuford, and M. J. Sepaniak, "Efficient disc on pillar substrates for surface enhanced Raman spectroscopy," Chemical Communications, vol. 47, no. 13, pp. 3814–3816, 2011. · ·

58. J.-S. Wi, E. S. Barnard, R. J. Wilson et al., "Sombrero-shaped plasmonic nanoparticles with molecular-level sensitivity and multifunctionality," ACS Nano, vol. 5, no. 8, pp. 6449–6457, 2011. · ·

59. M. J. Madou, Fundamentals of Microfabrication: The Science of Miniaturization, CRC, Boca Raton, Fla, USA, 2002.

60. E. Hao and G. C. Schatz, "Electromagnetic fields around silver nanoparticles and dimers," Journal of Chemical Physics, vol. 120, no. 1, pp. 357–366, 2004. · ·

61. M. Kahl, E. Voges, S. Kostrewa, C. Viets, and W. Hill, "Periodically structured metallic substrates for SERS," Sensors and Actuators B, vol. 51, no. 1–3, pp. 285–291, 1998.

62. K. R. Li, M. I. Stockman, and D. J. Bergman, "Self-similar chain of metal nanospheres as an efficient nanolens," Physical Review Letters, vol. 91, no. 22, pp. 227402/1–227402/4, 2003.

63. A. Polemi, S. M. Wells, N. V. Lavrik, M. J. Sepaniak, and K. L. Shuford, "Local field enhancement of pillar nanosurfaces for SERS," Journal of Physical Chemistry C, vol. 114, no. 42, pp. 18096–18102, 2010. · ·

64. H. Duan, A. I. Fernández-Domínguez, M. Bosman, S. A. Maier, and J. K. W. Yang, "Nanoplasmonics: classical down to the nanometer scale," Nano Letters, vol. 12, no. 3, pp. 1683–1689, 2012. · ·

65. H. Im, K. C. Bantz, N. C. Lindquist, C. L. Haynes, and S. H. Oh, "Vertically oriented sub-10-nm plasmonic nanogap arrays," Nano Letters, vol. 10, no. 6, pp. 2231–2236, 2010. · ·

66. M. R. Gartia, Z. Xu, E. Behymer et al., "Rigorous surface enhanced Raman spectral characterization of large-area high-uniformity silver-coated tapered silica nanopillar arrays," Nanotechnology, vol. 21, no. 39, Article ID 395701, 2010. · ·

67. J. Henzie, M. H. Lee, and T. W. Odom, "Multiscale patterning of plasmonic metamaterials," Nature Nanotechnology, vol. 2, no. 9, pp. 549–554, 2007. · ·

68. S. Aksu, M. Huang, A. Artar et al., "Flexible plasmonics on unconventional and nonplanar substrates,"Advanced Materials, vol. 23, no. 38, pp. 4422–4430, 2011. ·

69. K. Herman, L. Szabó, L. F. Leopold, V. Chiş, and N. Leopold, "In situ laser-induced photochemical silver substrate synthesis and sequential SERS detection in a flow cell," Analytical and Bioanalytical Chemistry, vol. 400, no. 3, pp. 815–820, 2011.

70. J. M. Yao, A. P. Le, M. V. Schulmerich et al., "Soft embossing of nanoscale optical and plasmonic structures in glass," ACS Nano, vol. 5, no. 7, pp. 5763–5774, 2011.

71. J. Kneipp, "Nanosenors based on SERS for application in living cells," in Surface-Enhanced Raman Scattering, K. Kneipp, M. Moskovits, and H. Kneipp, Eds., vol. 103, Springer, Heidelberg, Germany, 2006.

72. J. Kneipp, H. Kneipp, B. Wittig, and K. Kneipp, "Novel optical nanosensors for probing and imaging live cells," Nanomedicine, vol. 6, no. 2, pp. 214–226, 2010.

73. J. Kneipp, H. Kneipp, M. McLaughlin, D. Brown, and K. Kneipp, "In vivo molecular probing of cellular compartments with gold nanoparticles and nanoaggregates," Nano Letters, vol. 6, no. 10, pp. 2225–2231, 2006. · ·

74. K. Kneipp, A. S. Haka, H. Kneipp et al., "Surface-enhanced raman spectroscopy in single living cells using gold nanoparticles," Applied Spectroscopy, vol. 56, no. 2, pp. 150–154, 2002. · ·

75. J. Kneipp, H. Kneipp, W. L. Rice, and K. Kneipp, "Optical probes for biological applications based on surface-enhanced Raman scattering from indocyanine green on gold nanoparticles," Analytical Chemistry, vol. 77, no. 8, pp. 2381–2385, 2005.

76. K. Kneipp, H. Kneipp, I. Itzkan, R. R. Dasari, and M. S. Feld, "Ultrasensitive chemical analysis by Raman spectroscopy," Chemical Reviews, vol. 99, no. 10, pp. 2957–2975, 1999.

77. R. M. Jarvis, N. Law, I. T. Shadi, P. O›Brien, J. R. Lloyd, and R. Goodacre, "Surface-enhanced raman scattering from intracellular and extracellular bacterial locations," Analytical Chemistry, vol. 80, no. 17, pp. 6741–6746, 2008. · ·

78. M. A. Ochsenkühn, P. R. T. Jess, H. Stoquert, K. Dholakia, and C. J. Campbell, "Nanoshells for surface-enhanced raman

spectroscopy in eukaryotic cells: cellular response and sensor development," ACS Nano, vol. 3, no. 11, pp. 3613–3621, 2009.

79. S. W. Bishnoi, C. J. Rozell, C. S. Levin et al., "All-optical nanoscale pH meter," Nano Letters, vol. 6, no. 8, pp. 1687–1692, 2006. ·

80. K. L. Nowak-Lovato, B. S. Wilson, and K. D. Rector, "SERS nanosensors that report pH of endocytic compartments during FcεRI transit," Analytical and Bioanalytical Chemistry, vol. 398, no. 5, pp. 2019–2029, 2010. · ·

81. J. Kneipp, H. Kneipp, B. Wittig, and K. Kneipp, "Following the dynamics of pH in endosomes of live cells with SERS nanosensors," Journal of Physical Chemistry C, vol. 114, no. 16, pp. 7421–7426, 2010. · ·

82. J. P. Scaffidi, M. K. Gregas, V. Seewaldt, and T. Vo-Dinh, "SERS-based plasmonic nanobiosensing in single living cells," Analytical and Bioanalytical Chemistry, vol. 393, no. 4, pp. 1135–1141, 2009. · ·

83. C. E. Talley, L. Jusinski, C. W. Hollars, S. M. Lane, and T. Huser, "Intracellular pH sensors based on surface-enhanced raman scattering," Analytical Chemistry, vol. 76, no. 23, pp. 7064–7068, 2005. · ·

84. C. A. R. Auchinvole, P. Richardson, C. McGuinnes et al., "Monitoring intracellular redox potential changes using SERS nanosensors," ACS Nano, vol. 6, no. 1, pp. 888–896, 2012. · ·

85. H. Cho, B. R. Baker, S. Wachsmann-Hogiu et al., "Aptamer-based SERRS sensor for thrombin detection," Nano Letters, vol. 8, no. 12, pp. 4386–4390, 2008. · ·

86. N. Hamaguchi, A. Ellington, and M. Stanton, "Aptamer beacons for the direct detection of proteins,"Analytical Biochemistry, vol. 294, no. 2, pp. 126–131, 2001.

87. D. Saerens, L. Huang, K. Bonroy, and S. Muyldermans, "Antibody fragments as probe in biosensor development," Sensors, vol. 8, no. 8, pp. 4669–4686, 2008. ·

88. K. L. Brogan, K. N. Wolfe, P. A. Jones, and M. H. Schoenfisch, "Direct oriented immobilization of F(ab') antibody fragments on gold," Analytica Chimica Acta, vol. 496, no. 1-2, pp. 73–80, 2003. · ·

89. C. R. Yonzon, C. L. Haynes, X. Zhang, J. T. Walsh, and R. P. Van Duyne, "A glucose biosensor based on surface-enhanced Raman scattering: improved partition layer, temporal stability, reversibility, and resistance to serum protein interference," Analytical Chemistry, vol. 76, no. 1, pp. 78–85, 2004. · ·

90. H. Li, J. Sun, T. Alexander, and B. M. Cullum, "Implantable SERS nanosensors for pre-symptomatic detection of BW agents," in Chemical and Biological Sensing VI, Proceedings of SPIE, pp. 8–18, March 2005. · ·

91. H. Li, J. Sun, and B. M. Cullum, "Label-free detection of proteins using SERS-based immuno-nanosensors," Nanobiotechnology, vol. 2, no. 1-2, pp. 17–28, 2006. · ·

92. H. Li, J. Sun, and B. M. Cullum, "Nanosphere-based SERS immuno-sensors for protein analysis," inSmart Medical and Biomedical Sensor Technology II, Proceedings of SPIE, pp. 19–30, October 2004. · ·

93. M. Litorja, C. L. Haynes, A. J. Haes, T. R. Jensen, and R. P. Van Duyne, "Surface-enhanced Raman scattering detected temperature programmed desorption: optical properties, nanostructure, and stability of silver film over SiO_2 nanosphere surfaces," Journal of Physical Chemistry B, vol. 105, no. 29, pp. 6907–6915, 2001.

94. D. A. Stuart, J. M. Yuen, N. Shah et al., "In vivo glucose measurement by surface-enhanced Raman spectroscopy," Analytical Chemistry, vol. 78, no. 20, pp. 7211–7215, 2006. · ·

95. H. Li, J. Sun, and B. M. Cullum, "Label-free detection of proteins using SERS-based immuno-nanosensors," Nanobiotechnology, vol. 2, no. 1-2, pp. 17–28, 2006. · ·

96. E. Betzig, A. Lewis, A. Harootunian, M. Isaacson, and E. Kratschmer, "Near-field scanning optical microscopy (NSOM)—development and biophysical applications," Biophysical Journal, vol. 49, no. 1, pp. 269–279, 1986.

97. V. Deckert, D. Zeisel, R. Zenobi, and T. Vo-Dinh, "Near-field surface-enhanced Raman imaging of dye-labeled DNA with 100-nm resolution," Analytical Chemistry, vol. 70, no. 13, pp. 2646–2650, 1998.

98. D. Zeisel, V. Deckert, R. Zenobi, and T. Vo-Dinh, "Near-field surface-enhanced Raman spectroscopy of dye molecules adsorbed on silver island films," Chemical Physics Letters, vol. 283, no. 5-6, pp. 381–385, 1998.

99. B. Ren, G. Picardi, and B. Pettinger, "Preparation of gold tips suitable for tip-enhanced Raman spectroscopy and light emission by electrochemical etching," Review of Scientific Instruments, vol. 75, no. 4, pp. 837–841, 2004. · ·

100. U. Neugebauer, P. Rösch, M. Schmitt et al., "On the way to nanometer-sized information of the bacterial surface by tip-enhanced Raman spectroscopy," ChemPhysChem, vol. 7, no. 7, pp. 1428–1430, 2006. · ·

101. B. M. Cullum and M. E. Hankus, "SERS nanoimaging probes for characterizing extracellular surfaces," in Smart Biomedical and Physiological Sensor Technology V, Proceedings of SPIE, September 2007. · ·

102. M. E. Hankus and B. M. Cullum, "SERS probes for the detection and imaging of biochemical species on the nanoscale," in Smart Medical and Biomedical Sensor Technology IV, Proceedings of SPIE, October 2006. · ·

103. M. E. Hankus, G. Gibson, N. Chandrasekharan, and B. M. Cullum, "Surface-enhanced Raman scattering (SERS)—nanoimaging probes for biological analysis," in Smart Medical and Biomedical Sensor Technology II, Proceedings of SPIE, pp. 106–116, October 2004. · ·

104. M. E. Hankus, G. J. Gibson, and B. M. Cullum, "Characterization and optimization of novel surface-enhanced Raman scattering (SERS)-based nanoimaging probes for chemical imaging," in Smart Medical and Biomedical Sensor Technology III, Proceedings of SPIE, October 2005. · ·

105. B. M. Cullum and M. E. Hankus, "SERS nanoimaging probes for characterizing extracellular surfaces," in Smart Biomedical and Physiological Sensor Technology V, Proceedings of SPIE, September 2007. · ·

106. X. Yang, C. Gu, F. Qian, Y. Li, and J. Z. Zhang, "Highly sensitive detection of proteins and bacteria in aqueous solution using surface-enhanced raman scattering and optical fibers," Analytical Chemistry, vol. 83, no. 15, pp. 5888–5894, 2011. · ·

107. M. Çulha, M. Altunbek, S. Keskin, and D. Saatçi, "Manipulation of silver nanoparticles in a droplet for label-free detection of biological molecules using surface-enhanced Raman scattering," in Plasmonics in Biology and Medicine VIII, Proceedings of SPIE, January 2011. · ·

108. M. Kahraman, I. Sur, and M. Çulha, "Label-free detection of proteins from self-assembled protein-silver nanoparticle structures using surface-enhanced raman scattering," Analytical Chemistry, vol. 82, no. 18, pp. 7596–7602, 2010. · ·

109. Z. A. Combs, S. Chang, T. Clark, S. Singamaneni, K. D. Anderson, and V. V. Tsukruk, "Label-free raman mapping of surface distribution of protein A and IgG biomolecules," Langmuir, vol. 27, no. 6, pp. 3198–3205, 2011. · ·

110. S. Keskin, M. Kahraman, and M. Çulha, "Differential separation of protein mixtures using convective assembly and label-free detection with surface enhanced Raman scattering," Chemical Communications, vol. 47, no. 12, pp. 3424–3426, 2011. · ·

111. E. Papadopoulou and S. E. J. Bell, "Label-free detection of single-base mismatches in DNA by surface-enhanced raman spectroscopy," Angewandte Chemie, vol. 50, no. 39, pp. 9058–9061, 2011. · ·

112. D. Van Lierop, K. Faulds, and D. Graham, "Separation free DNA detection using surface enhanced raman scattering," Analytical Chemistry, vol. 83, no. 15, pp. 5817–5821, 2011. · ·

113. Z. Zhang, Y. Wen, Y. Ma, J. Luo, L. Jiang, and Y. Song, "Mixed DNA-functionalized nanoparticle probes for surface-enhanced Raman scattering-based multiplex DNA detection," Chemical Communications, vol. 47, no. 26, pp. 7407–7409, 2011. · ·

114. J. A. Dougan, D. MacRae, D. Graham, and K. Faulds, "DNA detection using enzymatic signal production and SERS," Chemical Communications, vol. 47, no. 16, pp. 4649–4651, 2011. · ·

115. J. D. Driskell and R. A. Tripp, "Label-free SERS detection of microRNA based on affinity for an unmodified silver nanorod array substrate," Chemical Communications, vol. 46, no. 19, pp. 3298–3300, 2010. · ·

116. L. Sun and J. Irudayaraj, "Quantitative surface-enhanced Raman for gene expression estimation," Biophysical Journal, vol. 96, no. 11, pp. 4709–4716, 2009.

117. S. Efrima and B. V. Bronk, "Silver colloids impregnating or coating bacteria," Journal of Physical Chemistry B, vol. 102, no. 31, pp. 5947–5950, 1998.

118. L. Zeiri, B. V. Bronk, Y. Shabtai, J. Eichler, and S. Efrima, "Surface-enhanced Raman spectroscopy as a tool for probing specific biochemical components in bacteria," Applied Spectroscopy, vol. 58, no. 1, pp. 33–40, 2004. · ·

119. L. Zeiri and S. Efrima, "Surface-enhanced Raman spectroscopy of bacteria: the effect of excitation wavelength and chemical modification of the colloidal milieu," Journal of Raman Spectroscopy, vol. 36, no. 6-7, pp. 667–675, 2005. · ·

120. A. Sengupta, M. L. Laucks, and E. James Davis, "Surface-enhanced Raman spectroscopy of bacteria and pollen," Applied Spectroscopy, vol. 59, no. 8, pp. 1016–1023, 2005. · ·

121. A. Sengupta, M. Mujacic, and E. J. Davis, "Detection of bacteria by surface-enhanced Raman spectroscopy," Analytical and Bioanalytical Chemistry, vol. 386, no. 5, pp. 1379–1386, 2006. · ·

122. M. L. Laucks, A. Sengupta, K. Junge, E. J. Davis, and B. D. Swanson, "Comparison of psychro-active arctic marine bacteria and common mesophillic bacteria using surface-enhanced raman spectroscopy,"Applied Spectroscopy, vol. 59, no. 10, pp. 1222–1228, 2005. · ·

123. R. M. Jarvis and R. Goodacre, "Discrimination of bacteria using surface-enhanced Raman spectroscopy," Analytical Chemistry, vol. 76, no. 1, pp. 40–47, 2004. · ·

124. R. M. Jarvis, A. Brooker, and R. Goodacre, "Surface-enhanced Raman spectroscopy for bacterial discrimination utilizing a scanning electron microscope with a Raman spectroscopy interface,"Analytical Chemistry, vol. 76, no. 17, pp. 5198–5202, 2004. · ·

125. W. R. Premasiri, D. T. Moir, M. S. Klempner, N. Krieger, G. Jones, and L. D. Ziegler, "Characterization of the surface enhanced Raman scattering (SERS) of bacteria," Journal of Physical Chemistry B, vol. 109, no. 1, pp. 312–320, 2005. ·

126. M. Kahraman, M. M. Yazıcı, F. Sahin, O. F. Bayrak, and M. Culha, "Reproducible surface-enhanced Raman scattering

spectra of bacteria on aggregated silver nanoparticles," Applied Spectroscopy, vol. 61, no. 5, pp. 479–485, 2007. · ·

127. I. Sayin, M. Kahraman, F. Sahin, D. Yurdakul, and M. Culha, "Characterization of yeast species using surface-enhanced Raman scattering," Applied spectroscopy, vol. 63, no. 11, pp. 1276–1282, 2009.

128. M. Culha, M. Kahraman, D. Çam, I. Sayin, and K. Keseroğlu, "Rapid identification of bacteria and yeast using surface-enhanced raman scattering," Surface and Interface Analysis, vol. 42, no. 6-7, pp. 462–465, 2010. · ·

129. A. Sujith, T. Itoh, H. Abe et al., "Imaging the cell wall of living single yeast cells using surface-enhanced raman spectroscopy," Analytical and Bioanalytical Chemistry, vol. 394, no. 7, pp. 1803–1809, 2009. · ·

130. P. Negri, A. Kage, A. Nitsche, D. Naumann, and R. A. Dluhy, "Detection of viral nucleoprotein binding to anti-influenza aptamers via SERS," Chemical Communications, vol. 47, no. 30, pp. 8635–8637, 2011. · ·

131. S. Shanmukh, L. Jones, J. Driskell, Y. Zhao, R. Dluhy, and R. A. Tripp, "Rapid and sensitive detection of respiratory virus molecular signatures using a silver nanorod array SERS substrate," Nano Letters, vol. 6, no. 11, pp. 2630–2636, 2006. · ·

132. S. D. Hudson and G. Chumanov, "Bioanalytical applications of SERS (surface-enhanced Raman spectroscopy)," Analytical and Bioanalytical Chemistry, vol. 394, no. 3, pp. 679–686, 2009. · ·

133. S. Xu, X. Ji, W. Xu et al., "Surface-enhanced Raman scattering studies on immunoassay," Journal of biomedical optics, vol. 10, no. 3, pp. 1–12, 2005.

134. M. Y. Sha, H. Xu, S. G. Penn, and R. Cromer, "SERS nanoparticles: a new optical detection modality for cancer diagnosis," Nanomedicine, vol. 2, no. 5, pp. 725–734, 2007. · ·

135. J. W. Chen, Y. Lei, X. J. Liu, J. H. Jiang, G. L. Shen, and R. Q. Yu, "Immunoassay using surface-enhanced Raman scattering based on aggregation of reporter-labeled immunogold nanoparticles.," Analytical and bioanalytical chemistry, vol. 392, no. 1-2, pp. 187–193, 2008.

136. C. C. Lin, Y. M. Yang, Y. F. Chen, T. S. Yang, and H. C. Chang, "A new protein A assay based on Raman reporter labeled immunogold nanoparticles," Biosensors and Bioelectronics, vol. 24, no. 2, pp. 178–183, 2008. · ·

137. H. Hwang, H. Chon, J. Choo, and J. K. Park, "Optoelectrofluidic sandwich immunoassays for detection of human tumor marker using surface-enhanced raman scattering," Analytical Chemistry, vol. 82, no. 18, pp. 7603–7610, 2010. · ·

138. M. Lee, S. Lee, J. H. Lee et al., "Highly reproducible immunoassay of cancer markers on a gold-patterned microarray chip using surface-enhanced Raman scattering imaging," Biosensors and Bioelectronics, vol. 26, no. 5, pp. 2135–2141, 2011. · ·

139. G. Wang, R. J. Lipert, M. Jain et al., "Detection of the potential pancreatic cancer marker MUC4 in serum using surface-enhanced Raman scattering," Analytical Chemistry, vol. 83, no. 7, pp. 2554–2561, 2011. · ·

140. E. Temur, I. H. Boyacl, U. Tamer, H. Unsal, and N. Aydogan, "A highly sensitive detection platform based on surface-enhanced Raman scattering for Escherichia coli enumeration," Analytical and Bioanalytical Chemistry, vol. 397, no. 4, pp. 1595–1604, 2010. · ·

141. B. Guven, N. Basaran-Akgul, E. Temur, U. Tamer, and I. H. Boyaci, "SERS-based sandwich immunoassay using antibody coated magnetic nanoparticles for Escherichia coli enumeration," Analyst, vol. 136, no. 4, pp. 740–748, 2011. · ·

142. X. X. Han, B. Zhao, and Y. Ozaki, "Surface-enhanced Raman scattering for protein detection," Analytical and Bioanalytical Chemistry, vol. 394, no. 7, pp. 1719–1727, 2009. · ·

143. Z. Zhou, G. G. Huang, and Y. Ozaki, "Label-free rapid semiquantitative detection of proteins down to sub-monolayer coverage by using surface-enhanced raman scattering of nitrate ion," Chemistry Letters, vol. 39, no. 11, pp. 1203–1205, 2010. · ·

144. I. S. Patel, W. R. Premasiri, D. T. Moir, and L. D. Ziegler, "Barcoding bacterial cells: a SERS-based methodology for pathogen identification," Journal of Raman Spectroscopy, vol. 39, no. 11, pp. 1660–1672, 2008. · ·

145. M. Kahraman, M. M. Yazici, F. Şahin, and M. Çulha, "Experimental parameters influencing surface-enhanced Raman scattering of bacteria," Journal of Biomedical Optics, vol. 12, no. 5, Article ID 054015, 2007. · ·

146. M. Kahraman, M. M. Yazici, F. Şahin, and M. Çulha, "Convective assembly of bacteria for surface-enhanced Raman scattering," Langmuir, vol. 24, no. 3, pp. 894–901, 2008. · ·

147. M. Kahraman, K. Keseroğlu, and M. Çulha, "On sample preparation for surface-enhanced Raman scattering (SERS) of bacteria and the source of spectral features of the spectra," Applied Spectroscopy, vol. 65, no. 5, pp. 500–506, 2011.

148. W. R. Premasiri, Y. Gebregziabher, and L. D. Ziegler, "On the difference between surface-enhanced Raman scattering (SERS) spectra of cell growth media and whole bacterial cells," Applied Spectroscopy, vol. 65, no. 5, pp. 493–499, 2011. · ·

149. D. Bhandari, S. M. Wells, A. Polemi, I. I. Kravchenko, K. L. Shuford, and M. J. Sepaniak, "Stamping plasmonic nanoarrays on SERS-supporting platforms," Journal of Raman Spectroscopy, 2011. · ·

150. A. J. Pasquale, B. M. Reinhard, and L. Dal Negro, "Engineering photonic-plasmonic coupling in metal nanoparticle necklaces," ACS Nano, vol. 5, no. 8, pp. 6578–6585, 2011. · ·

151. M. Kahraman, N. Tokman, and M. Çulha, "Silver nanoparticle thin films with nanocavities for surface-enhanced Raman scattering," ChemPhysChem, vol. 9, no. 6, pp. 902–910, 2008.

Chapter 3

Polarization Dependence of Surface Enhanced Raman Scattering on a Single Dielectric Nanowire

Hua Qi, R. W. Rendell, O. J. Glembocki,
and S. M. Prokes

Electronics Science and Technology Division, Naval Research Laboratory, Washington, DC 20375, USA

ABSTRACT

Our measurements of surface enhanced Raman scattering (SERS) on Ga_2O_3 dielectric nanowires (NWs) core/silver composites indicate that the SERS enhancement is highly dependent on the polarization direction of the incident laser light. The polarization dependence of the SERS signal with respect to the direction of a single NW was studied by changing the incident light angle. Further investigations demonstrate that the SERS intensity is not only dependent on the direction and wavelength of the incident light, but also on the species of the SERS active molecule. The largest signals were observed on an NW when the incident 514.5 nm light was polarized perpendicular to the length of the NW, while the opposite phenomenon was observed at the wavelength of 785 nm. Our theoretical simulations of the polarization dependence at 514.5 nm and 785 nm are in good agreement with the experimental results.

INTRODUCTION

Surface enhanced Raman scattering (SERS) has been regarded as a unique technique to detect trace levels of chemical compounds, since the vibrational information is very specific to the bonds in the molecules. In recent years, there has been significant interest in exploring various nanostructures as SERS substrates to optimize the electromagnetic field enhancement and significantly improve sensitivity. In the case of nonspherical nanoparticles it has been shown that the SERS enhancement is strongly dependent on the direction of polarization of the exciting incident light [1–12]. Of particular interest are cylindrical geometries because the enhancement is a function of the aspect ratio. In addition, in the case of very long Ag nanowires it has been shown that surface plasmons can be launched at one end of the nanowires and travel to the other end. Recently there has been interest in metal coated dielectric core nanowires because they form plasmonic shells [13, 14]. These geometries can form the basis of plasmonically modulated photonic devices. It is well known that the local electric field in plasmonic coupling of closely spaced particles, usually called "hot spots," can be orders of magnitude stronger than those on individual particles. The importance of "hot spots" in SERS process has been widely discussed [15–21].

In this work, a highly effective SERS composite of dielectric Ga_2O_3 NWs core/silver was employed to investigate the SERS intensity dependence on the laser polarization. Experimental results show that both variable wavelength and angle of the incident light play very important roles on the resulting SERS intensity. Our theoretical simulations indicated that the maximum SERS enhancement could be obtained when the polarization is perpendicular to the NW length at 514.5 nm excitation, while the opposite phenomena should be expected at a laser wavelength of 785 nm. All of these expectations are in good agreement with our experimental observations. In addition, further experiment shows that the orientation of self-assembled monolayer of active SERS molecules on the NWs may affect the SERS enhancement as well.

EXPERIMENTAL DETAILS

Random Ga_2O_3 NWs were grown via the vapor-liquid-solid (VLS) growth mechanism [22]. The gallium metal (99.995% pure) used as a source was placed 6 inches upstream from the Si (100) substrate, which had a 20 nm gold film. The furnace was heated to 900°C while flowing simultaneously a mixture of argon and oxygen gases in a ratio of 6 : 1 through the tube. To study the SERS behavior of a single NW, the Ga_2O_3 NWs as grown on a Si substrate were sonicated off in methanol, dropped and dried on a bare silicon surface for further SERS study.

The discrete single NWs were covered with a layer of silver produced by an electroless (EL) plating approach for the SERS behavior study, which has been described in our previous report [13, 14]. Briefly, the silver ion in Tollen reagent was reduced to neutral Ag and then homogenously deposited on the NWs surface. The morphology of the silver on the NW surface is directly related to the chemical concentration and reaction time. Usually, short reaction time of the silver electroless plating process results in closely spaced but separate nanoparticles, while longer times lead to linking up of the islands into a rough layer.

The SERS line maps were carried out using a confocal μ-Raman system which consisted of a Mitutoyo Microscope and an Ocean Optics QE65000 spectrometer equipped with a thermoelectrically cooled CCD. The 514.5 nm line of an Ar+ ion laser was used as the excitation source. The microscope utilized a 100x 0.7 NA objective for focusing the laser light and was coupled to the spectrometer through a fiber optic cable. The maps were collected with low laser power of 0.75 mW at the sample. This was done to prevent desorption and damage to the benzenethiol and to prevent alterations to the Ag layer. The SERS intensity dependence on variable polarization angles was performed using a Delta Nu system which consists of an Olympus Microscope and a Raman spectrometer equipped with a thermoelectrically cooled

CCD. The 785 nm line of Ti: Sapphire laser was used as the excitation source to detect the SERS strength dependence on a single NW/silver composite. The microscope utilized a 50x 0.75 NA objective for focusing the laser light. The spectra were collected with a laser power of 3 mW at the sample.

The full Maxwell's equations are solved numerically using the finite elements method (FEM) through the RF module of the COMSOL Multiphysics finite-elements package [23]. We used FEM simulations because they are more appropriate for the complex geometries of nanowires placed on a substrate [24]. FEM approaches have been shown to agree well with finite difference time domain simulations [24, 25]. In addition, the FEM technique implemented via COMSOL has been shown to be capable of accurate simulation for nanoshell structures on substrates if sufficiently accurate meshing is used and appropriate absorbing perfectly matched layers PML are used on the boundaries [26–28]. This approach allows us to treat the scattering problem by bringing in planes waves of a specific polarization and then calculating the scattered fields. As another test of the numerical simulations, we find excellent agreement between the COMSOL results with the analytical Mie solutions for scattering from an Ag sphere [24, 27]. The modeling of the nanowires in air above an Si substrate requires the construction of separate PML boundaries for both the Si and air regions. The dielectric constants of each PML match its adjoining region, either Si or air. To avoid artifacts due to scattering at the boundaries between the Si and air regions of the PML's, the scattering at the air-Si interface is incorporated into the incident-polarized plane wave by means of analytical functions for the Fresnel coefficients. The resulting incident field is thus an exact solution in the absence of the nanowire. The numerical finite element solution for the scattered field is then calculated in the presence of the nanowire. This requires sufficient meshing to adequately resolve both the nanowires with shells and the PML boundaries [27, 28].

RESULTS AND DISCUSSION

To demonstrate the surface enhanced Raman scattering (SERS) polarization dependence on the direction of the nanowires (NWs), we carried out the mapping investigations on NW/Ag composite on the same NW using a 514.5 nm excitation with parallel and perpendicular laser polarization, respectively. The Ga_2O_3 NWs, used in all experiments, were grown via the vapor-liquid-solid (VLS) growth mechanism. Empirically, the length and diameter of Ga_2O_3 NWs depended on the growth time and gas flow rate, as well as the catalyst size. In our growth, the diameters varied from 50 nm to 300 nm, and the length was usually greater than 10 microns. Energy dispersive X-ray (EDX) analysis showed that the chemical composition of the NWs was stoichiometric Ga_2O_3. Benzenethiol (BT) was used as the SERS active molecules.

The intensity dependence maps of the 1576 cm^{-1} BT SERS band on incident laser polarization are shown in Figures 1(a) and 1(b), by which the polarization dependence can be easily visualized. Both of the images were mapped at the same region of the sample by rotating the sample. Figure 1(c) shows the corresponding microscope image of the SERS mapping area. From the maps, it is obvious that the signal strength is maximized when the laser is perpendicular to the NW.

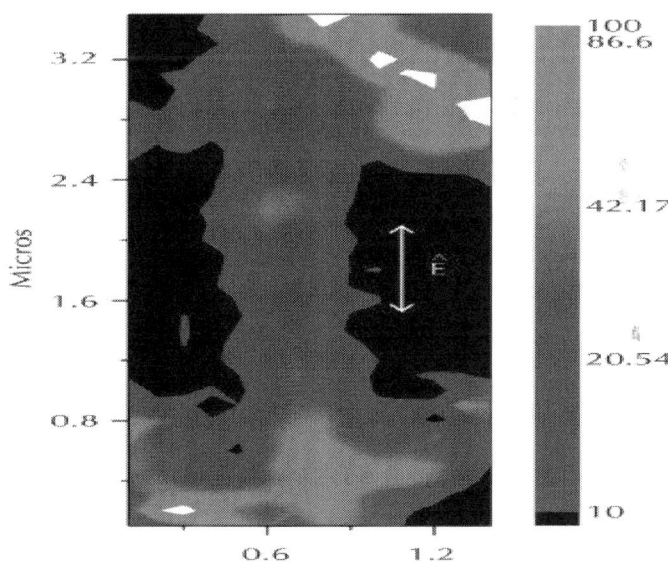

(a)

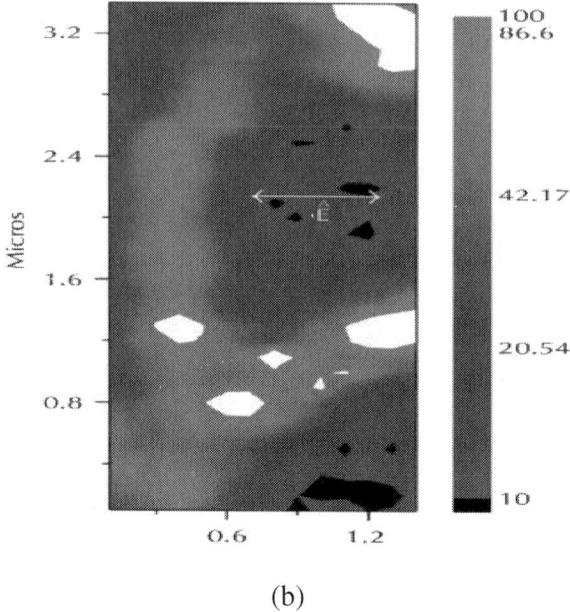

(b)

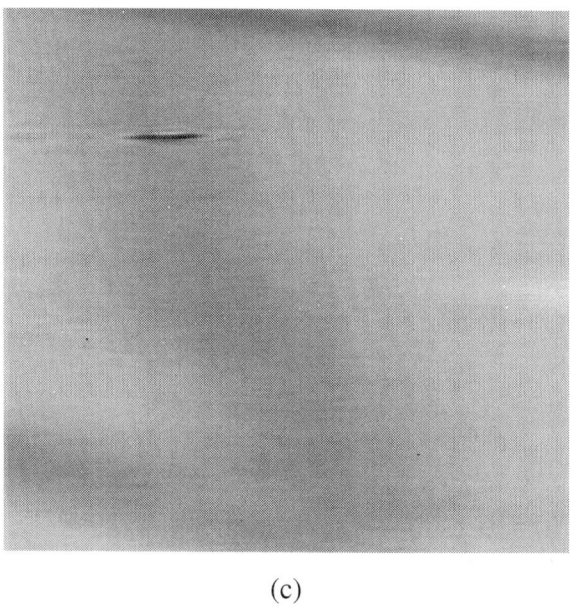

(c)

Figure 1: Experimental maps of the 1576 cm⁻¹ BT SERS band of NW/Ag composite using a laser with 514.5 nm wavelength at the same NW, and the sample was rotated for parallel/perpendicular polarization mapping; the polarization is parallel (a) and perpendicular (b) to the NW. (c) Corresponding microscope image of the SERS mapping region of (a) and (b).

To verify our experimental observation, we performed the simulations using the RF module of COMSOL Multiphysics, which provides a finite-element solution of Maxwell's equations. Figures 2(a) and 2(b) show the simulated SERS maps of the NW with length of 500 nm at a laser wavelength of 514.5 nm, indicating that the strongest SERS signal occurs when the polarization is perpendicular to the NW longitude, which clearly confirmed the incident laser polarization effects on the SERS strength, as we obtained in experiments shown in Figure 1. It is noted in the maps that the plasmon oscillations are clearly observed along the NW when the polarization is parallel to the NW, while the transverse modes have excitations along the radial direction and hence much shorter period and uniform in NW's longitudinal direction.

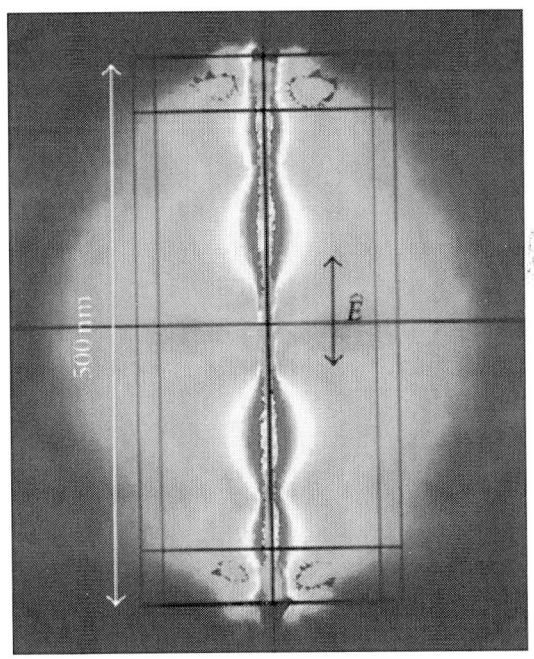

(a)

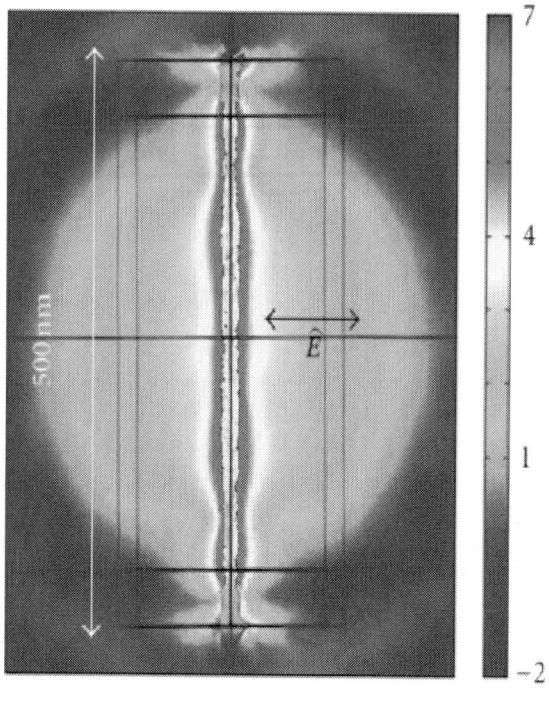

(b)

Figure 2: Log plots of SERS enhancement simulated at a wavelength of 514.5 nm when the NW direction is parallel (a) and perpendicular (b) to the polarization.

Furthermore, more comprehensive simulations were carried out in the range of wavelength from 400 nm to 900 nm, as shown in Figure 3, which includes the wavelength of 514.5 nm used in the mapping experiments and simulations above. By comparison of Figures 3(a) and 3(b), it is obvious that, in the case of parallel, the SERS enhancement factor is $10^6 \sim 10^7$, while the perpendicular case shows an enhancement factor of $10^8 \sim 10^9$, clearly indicating that the NW/Ag composites show stronger SERS enhancement when the laser polarization is perpendicular to the NW direction at 514.5 nm, which is in good agreement with the experimental mapping images of Figure 1 and the theoretical simulation mapping results of Figure 2.

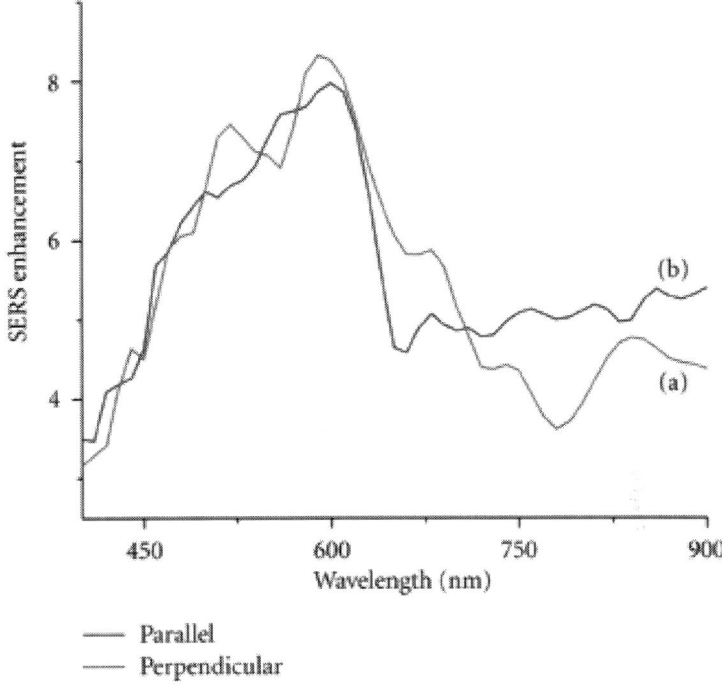

— Parallel
— Perpendicular

Figure 3: Simulations of SERS enhancements in the range of wavelength from 400 nm to 900 nm. The NW direction is parallel (red) and perpendicular (blue) to the polarization.

From Figures 3(a) and 3(b), it is expected, at the wavelength of 785 nm (red laser), that the strongest SERS signals should be observed when the laser polarization parallelized to the NW longitude direction, which is completely opposite to the observations at the excitation of 514.5 nm. Hence, we performed further investigations of SERS enhancements dependence on the incident angles using incident laser of 785 nm and 514.5 nm. In the process of measurements, data were collected at three different angles, including 0°, 45°, and 90°, respectively. The representative BT SERS line of 1576 cm^{-1} at different angles and laser wavelengths are shown in Figure 4. The red and green lines are roughly linear fitting of the experimental data obtained at three angles, which clearly indicates the trends of SERS enhancement dependence on the incident laser angles and the different wavelengths. It is noted that the data were normalized to the maximum intensity.

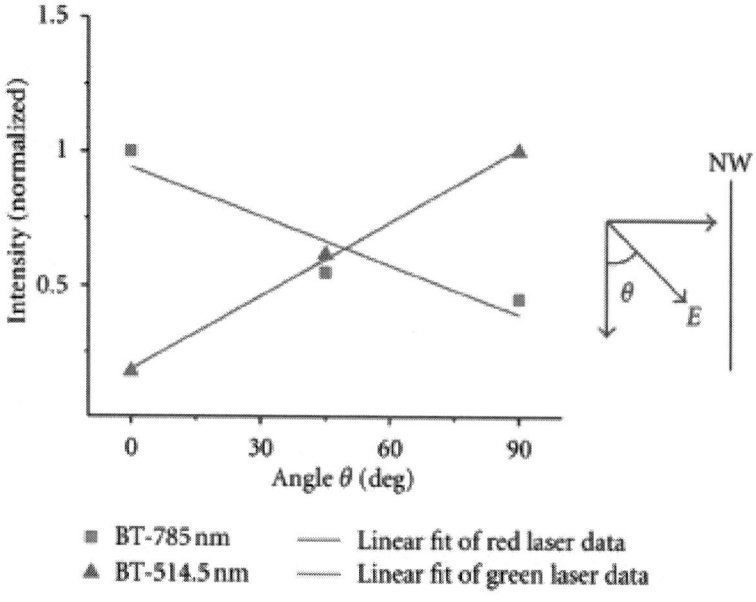

Figure 4: Experimental polarization angle dependence of surface enhanced Raman scattering (SERS) on nanowire at the wavelength of λ= 785 nm (red squares and its linear fitting line) and λ= 514.5 nm (green triangles and its linear fitting line) the data were normalized to the maximum intensity.

More detailed simulations of the SERS polarization dependence of the NW core/Ag composites for the incident wavelength of 785 nm and 514.5 nm are shown in Figures 5(a) and 5(b), respectively, which provide a continuous SERS enhancement trend from 0° to 90°. As can be seen in the calculation, the largest SERS enhancement is completely flipped at the excitation wavelength of 785 nm and 514.5 nm, which matches very well with our experimental observations as shown in Figure 4. For SERS, the Raman intensity generally increases by a factor of E^4, where the first two powers of the enhancement are due to the local electromagnetic field which in the present case is predominantly due to the contact between the nanowire and the Si substrate. The second two powers of the enhancement factor are the Raman emission enhancement. As a result of these two contributions to the SERS enhancement, it has been argued [9] that this should lead to a simple $\sin^2(\theta)$ polarization dependence for a nanowire. If θ is the polarization angle of the incident light with respect to the direction along the length of the nanowire, then the local electric field is $|E_{loc}(\omega, \theta)| = |E_{max}(\omega)| \sin(\theta)$ where E_{max} is the maximum of the local electric field. However, the direction of the induced electric field between the nanowire and the substrate is always in the direction across the nanowire-substrate contact $(\theta = \pi/2)$ and independent of the incident polarization. Thus the total SERS enhancement is expected to be proportional towhere generally a phase shift phase is also expected due to the fact that metallic nanowires on substrates which are tens of microns in length can function as Fabry-Perot cavities resulting in geometry specific standing plasmon waves [26, 29, 30].

$$G = \left| \frac{E_{max}(\omega_L)}{E_0(\omega_L)} \right|^2 \left| \frac{E_{max}(\omega_R)}{E_0(\omega_R)} \right|^2 \sin^2(\theta + \varphi_{phase})$$

$$= \left| \frac{E_{max}}{E_0} \right|^4 \sin^2(\theta + \varphi_{phase}),$$

(1)

Polarization dependences as in (1) have been previously observed [9, 30]. This is also consistent with the simulation results in Figure 5 where the difference between (a) and (b) can be described in terms of a different phase shift occurring at each excitation wavelength. This can be expected due to the observations that in the Fabry-Perot behavior of nanowires on substrates [26, 29], the surface plasmon wavelength has been found to differ from the excitation wavelength (e.g., between 400–600 nm for excitation at 785 nm [26, 29]) and is geometry dependent. As a result, the plasmon standing waves will generally occur with different relative phase shifts and the maximum enhancement may occur at a different polarization angle depending on the excitation wavelength. This has been observed for the first time in the present work and the result is consistent with previous observations of metallic nanowires on substrates [26, 29, 30].

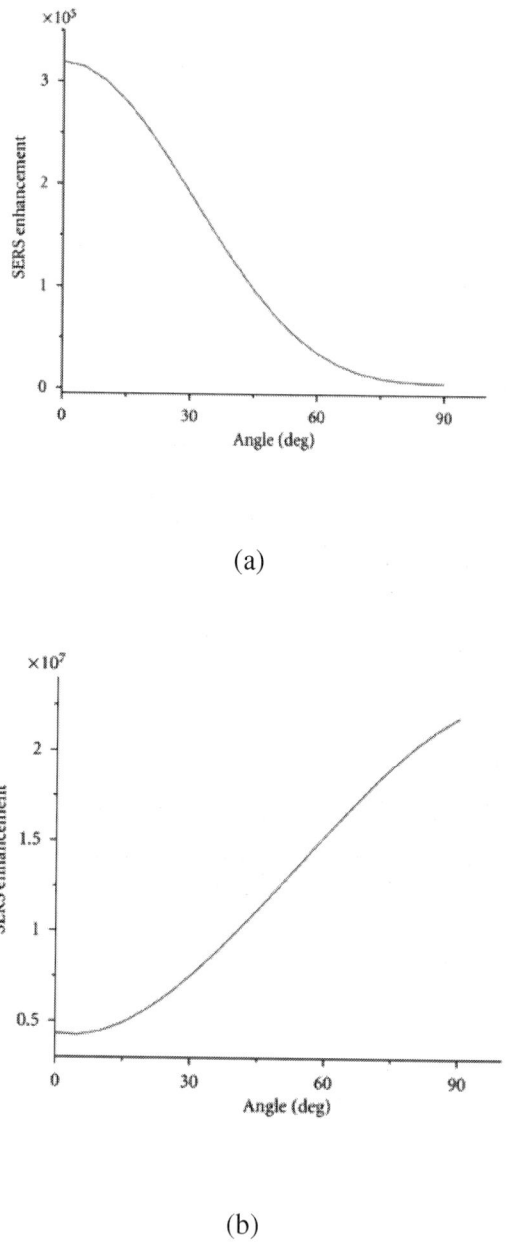

(a)

(b)

Figure 5: Simulation of polarization angle dependence of surface enhanced Raman scattering (SERS) averaged over nanowire surface excluding the wire ends for (a) $\lambda = 785$ nm and (b) $\lambda = 514.5$ nm.

From the comprehensive simulation results of Figure 3, we can see that the polarization angle dependence of the wire at 633 nm would exhibit similar behavior to that shown for the 514 nm in Figure 5(b), but the SERS enhancement in the 90° case shows stronger enhancements and more extended fields compared to the 514 nm case shown in Figure 2(a). For the 0° case, the SERS enhancement would be predicted to be weaker at 633 nm compared to the 514 nm case. This is shown in Figure 6.

633 nm

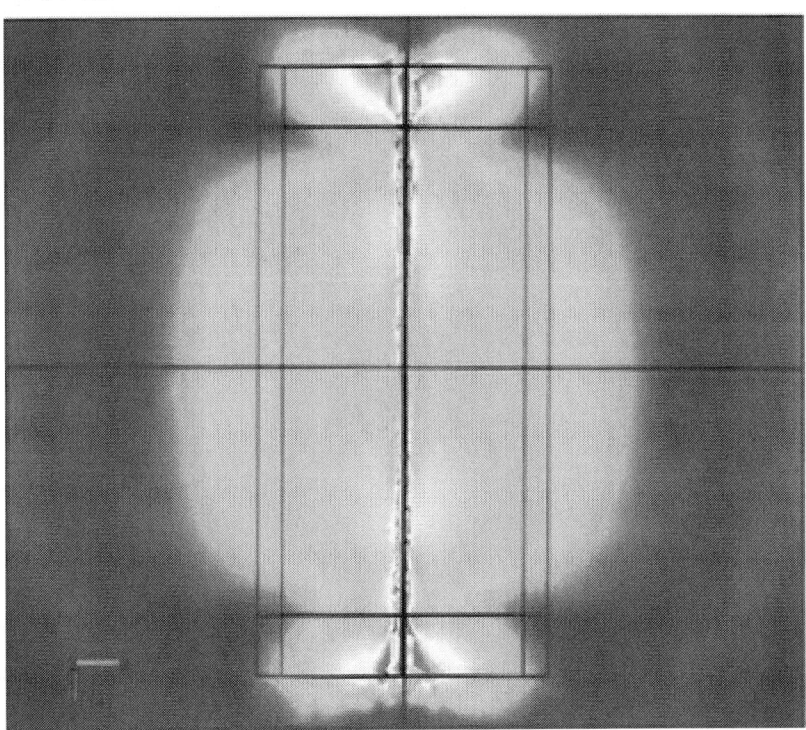

0°

(a)

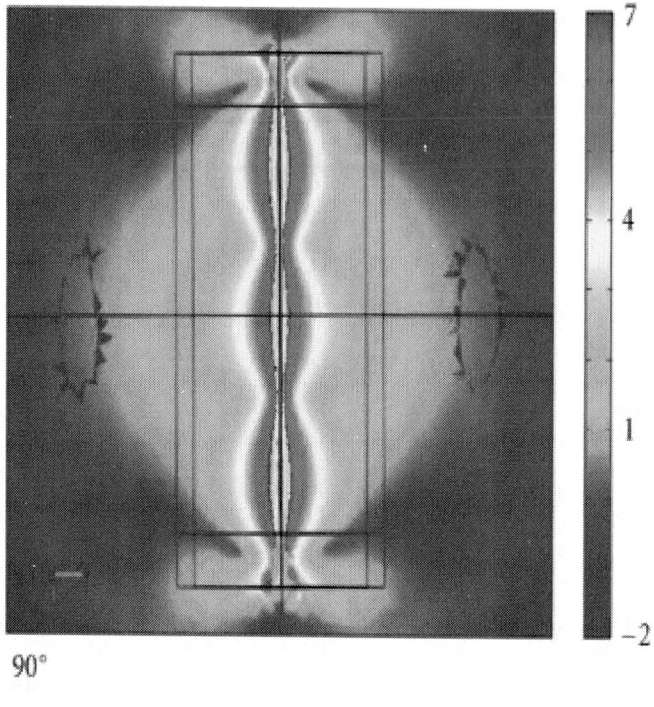

90°

(b)

Figure 6: Log plots of SERS enhancement simulated at a wavelength of 633 nm when the NW direction is parallel (a) and perpendicular (b) to the polarization.

In addition, the NW/Ag composites SERS dependence on the polarization was experimentally investigated using a laser excitation of 785 nm to investigate the incident angle effects on the SERS enhancement in detail. Figures 7(a) and 7(b) show the high resolution SEM image of an isolated bare NW and NW/Ag composites. As shown in Figure 7(b), a rough layer of silver was clearly plated on the NW surface, which is produced by an electroless (EL) approach [13, 14]. From the SEM image, the silver nanoparticles display 3D islands with a diameter range of 20 nm. The space between the particles is less than 10–20 nm, which leads to the strong plasmonic oscillation due to coupling of individual nanoparticles/island formed on NWs, as we reported previously [13]. The SERS spectra were carried out by changing the angle θ in 15° increments between the electric field of the incident light and the axis parallel to the NW length, as shown in the sketch of Figure 7(b). To minimize the time-dependence degradation influence of the SERS signal intensity during the measurements, two cycles of SERS measurements, from 0° to 90°, then back to 0°, were performed on a single NW. The data obtained from those two cycles were averaged for the spectra plots of each angle.

(a)

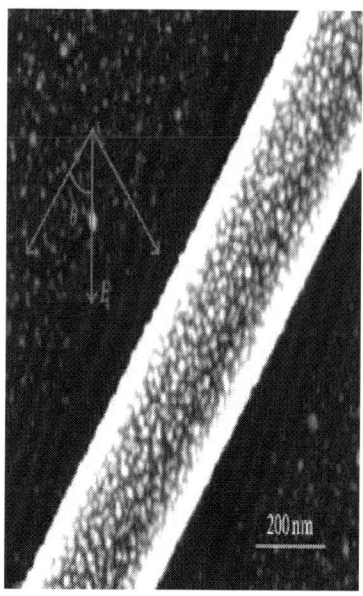

(b)

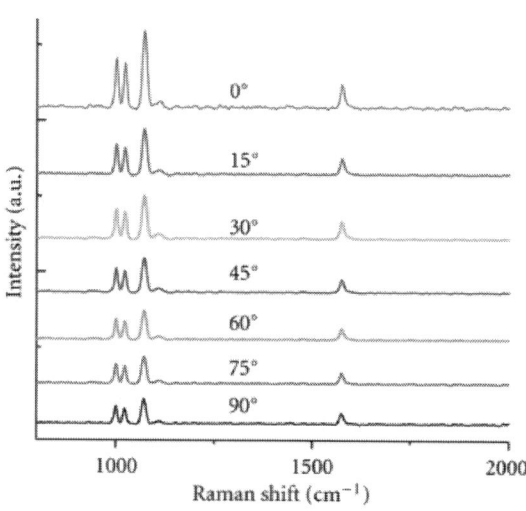

(c)

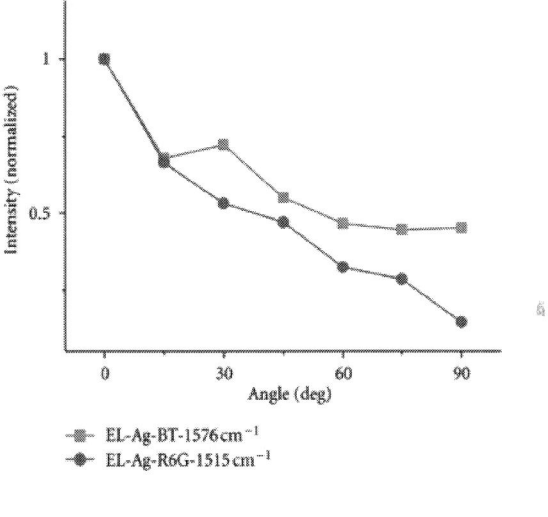

(d)

(e)

Figure 7: SEM image of a bare (a) and an electroless silver plated (b) NW. Inset of (b): the sketch indicates the laser polarization direction. (c) SERS dependence of a single NW/Ag composite on the laser polarization angles for Benzenethiol (BT) molecules at a wavelength of 785 nm. (d) Plot of intensity of SERS representative lines of BT (red) and rhodamine 6G (R6G) (blue) versus angles and the data were normalized to the maximum intensity. (e) SERS dependence of a single NW/Ag composite on the laser polarization angles for R6G molecules at a wavelength of 785 nm.

Figure 7(c) shows a series of SERS spectra as a function of the incident light polarization angles from 0° to 90°. Here benzenethiol (BT) molecules were used as SERS active molecules. The major Raman peaks at 1001, 1023, 1073, and 1574 cm^{-1} can be assigned to symmetric ring breathing, in-plane phenyl ring bending, in plane C–H bending, and in plane C–C stretching of the phenyl ring from BT, respectively, which is in good agreement with those reported previously [31–36]. It is clear that the SERS signal displayed strong dependence on the polarization direction of the incident laser on a single NW, and the strongest SERS intensity was observed when the polarization is parallel to the NW length at an angle of 0°.

A plot of the normalized intensity data of the representative BT SERS band as a function of polarization angle is shown in Figure 7(d) (red curve), by which the polarization dependence can be easily visualized. It is obvious that the strength of SERS signals is maximized at 0°, which is different from the SERS strength dependence on laser polarization reported previously on pure silver or gold NWs with sparse noble metal particles [8, 9]. The latter reported that the maximum of the SERS enhancement was observed when the polarization was perpendicular to the metal NWs. The difference may be attributed to the fact that the dielectric Ga_2O_3 NWs SERS substrate and the uniform coverage of silver particles layer play important roles in this polarization dependence process. That is to say the NWs SERS behavior is also related to the structure and composites of the SERS substrate.

Additionally, a bump was observed at the angle of around 30° for BT molecules as shown in Figure 7(d) (red curve). A possible explanation to this observed bump around 30° is that the orientation of the attached active molecules may have effect on this polarization dependence behavior. It is well known that the benzene rings of BT could easily form a self-assembled monolayer on a silver surface. In our study, the dielectric core was wrapped with a layer of silver, which provides a favorable condition for the formation of the ordered self-assembled monolayer (SAM). However, numerous reports in the literatures indicate that the orientation of the phenyl ring plane vary from perpendicular to flat from the substrate surface [37–45]. The SAM orientation of BT may affect the SERS enhancement factor when the incident light angle is variable. Here, as shown in the Figure 8(a), we assumed that the BT SAM was deduced to be tilted about 60° from the substrate surface, which is similar to the research results obtained by Szafranski et al. [37]. Hence the abrupt increase of SERS signals at the angle of around 30° could be attributed to the perpendicularity between the laser polarization direction and the phenyl ring plane. This is consistent with the electrostatic model interpretation of Gunnarsson et al. [46] for the relative SERS intensities of rhodamine 6G and thiophenol in terms of the orientation of the phenol ring out from the surface. This observation implies that the SERS signal strength highly depends on the molecules packing structure on the active substrate, which may provide a useful way to investigate the orientation of self-assembly molecular monolayer on surface via SERS technique.

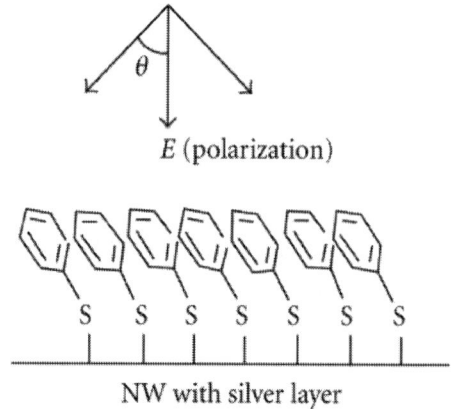

(a)

(b)

Figure 8: (a) A scheme of BT molecules orientation on the substrate and their interaction with polarized laser. (b) The structure of R6G.

To eliminate the effect of the SAM on the polarization behavior, and to investigate the pure polarization dependence on the NW direction, a random physically adsorbed layer of rhodamine 6G (R6G) was employed to study the SERS dependence on a single Ga_2O_3 NW. Figure 7(e) shows a series of SERS spectra of R6G with variable angles. Since R6G was attached randomly on an NW surface, this could effectively remove the polarization effect caused by the ordered molecular orientation. Hence the SERS polarization dependence behavior would completely come from the NW/Ag composite. Representative SERS line of R6G versus angle was plotted in Figure 7(d), which clearly show that the signal intensity of R6G increases with a decrease of the incident light angle. The signal is maximized when the polarization direction of the light is parallel to the NW direction ($\theta = 0°$), which is in good agreement with the observation of BT molecule at the same incident wavelength of 785 nm.

It is observed that R6G random molecules layer shows a stronger polarization effect than BT ordered layer, which is opposite to what one would expect. However if the size of the molecules and the roughness of the silver particles on NWs are taken into account, the fact should be easier to understand. As shown in Figure 8, it is clear to see that the area of R6G is much larger than BT. This means that, per unit, the number of R6G molecules is much smaller than BT, which results in the ordered fact of benzene rings in the case of R6G, somehow. Additionally the silver particles on NWs are separated island, not perfectly smooth surface, which may cause the disorder pack of the BT molecules, and reduce the enhancement dependence on the polarization. Based on these facts, it is possible and reasonable that R6G layer shows a stronger polarization effect.

It has been reported that the bare single-crystalline silver NW showed SERS enhancement only when the laser polarization direction was perpendicular to the long axis of the NW [47]. In the case of the dielectric Ga_2O_3NW core/silver layer composites, we observed SERS enhancements for both perpendicular and parallel polarization directions. This can also be attributed to the structure difference of the SERS active substrate. Some calculations indicated that the maximum enhancement of the plasmon resonances in noble metal spheroids arises from the region of the highest curvature [48–50]. Recently Hill et al. reported the high efficacy of gold-nanoparticle- (NP-) gold film system in generating large field enhancements due to the extremely close distance between the NP and the film, clearly demonstrating the importance of the substrate gap for SERS enhancement [51]. We have also demonstrated that there is a significant interaction between the spherical silver NPs on the ZnO NW due the closely spaced NPs, and the islands also interact with the dielectric NW substrate, further increasing the enhancement [13]. In addition, the SERS enhancement dependence on the polarization direction was observed for all bands at the excitation source of 785 nm and 514.5 nm. In recent years, a significant amount of work has been published on SERS, some of which has been on nanostructures, such as 1D nanowires [52, 53]. However, there are no reports of results such as those that we are presenting in this work. Thus all of these studies deepen the understanding of the strong SERS enhancement using dielectric NW core/silver layer composites and the underlying plasmonic behavior, which is critical to the development of sensor technology that can eventually be used for trace chemical, biological, and explosive sensing.

CONCLUSIONS

We have demonstrated the strong SERS dependence on the polarization direction of the incident light by investigating dielectric Ga_2O_3 NW core/silver composites, which showed that the SERS enhancement not only relies on the polarization direction of light with respect to the NW, but also on the wavelength of the incident laser and the orientation of SERS active molecules. Specifically, the strongest intensity of the SERS enhancement is obtained when the polarization of the incident light is parallel to the NW length at the incident wavelength of 785 nm, while the signal is maximized when the polarization is perpendicular to the NW length at 514.5 nm, which were confirmed by both experimental and theoretical simulation results. Therefore the SERS enhancement can be maximized by optimizing the direction of the NW, the incident angle of the laser polarization, the orientation of the molecules, and the wavelength of the laser.

ACKNOWLEDGMENTS

This work was partially supported by the Office of Naval Research (ONR) and Nanoscience Institute (NSI) of the US Naval Research Laboratory. The authors thank Dr. D. A. Alexson for his help during some data collections.

REFERENCES

1. H. Xux and M. Käll, "Polarization-dependent surface-enhanced raman spectroscopy of isolated silver nanoaggregates," ChemPhysChem, vol. 4, no. 9, pp. 1001–1005, 2003. · ·

2. J. P. Kottmann and O. J. F. Martin, "Plasmon resonant coupling in metallic nanowires," Optics Express, vol. 8, no. 12, pp. 655–663, 2001.

3. S. A. Maier, P. G. Kik, H. A. Atwater et al., "Local detection of electromagnetic energy transport below the diffraction limit in metal nanoparticle plasmon waveguides," Nature Materials, vol. 2, no. 4, pp. 229–232, 2003. · ·

4. T. Dadosh, J. Sperling, G. W. Bryant et al., "Plasmonic control of the shape of the raman spectrum of a single molecule in a silver nanoparticle dimer," ACS Nano, vol. 3, no. 7, pp. 1988–1994, 2009. · ·

5. I. K. J. Kretzers, R. J. Parker, R. V. Olkhov, and A. M. Shaw, "Aggregation kinetics of gold nanoparticles at the silica-water interface," Journal of Physical Chemistry C, vol. 113, no. 14, pp. 5514–5519, 2009. · ·

6. Y. Sun, G. Wei, Y. Song et al., "Type I collagen-templated assembly of silver nanoparticles and their application in surface-enhanced Raman scattering," Nanotechnology, vol. 19, no. 11, Article ID 115604, 2008. · ·

7. R. Gunawidjaja, S. Peleshanko, H. Ko, and V. V. Tsukruk, "Bimetallic nanocobs: decorating silver nanowires with gold nanoparticles," Advanced Materials, vol. 20, no. 8, pp. 1544–1549, 2008. · ·

8. J. L. Seung, M. B. Jeong, and M. Moskovits, "Polarization-dependent surface-enhanced raman scattering from a silver-nanoparticle-decorated single silver nanowire," Nano Letters, vol. 8, no. 10, pp. 3244–3247, 2008. · ·

9.   H. Wei, F. Hao, Y. Huang, W. Wang, P. Nordlander, and H. Xu, "Polarization dependence of surface-enhanced Raman scattering in gold nanoparticle-nanowire systems," Nano Letters, vol. 8, no. 8, pp. 2497–2502, 2008. · ·

10.   B. J. Wiley, Y. Chen, J. M. McLellan et al., "Synthesis and optical properties of silver nanobars and nanorice," Nano Letters, vol. 7, no. 4, pp. 1032–1036, 2007. ·

11.   J. M. McLellan, Z. Y. Li, A. R. Siekkinen, and Y. Xia, "The SERS activity of a supported ag nanocube strongly depends on its orientation relative to laser polarization," Nano Letters, vol. 7, no. 4, pp. 1013–1017, 2007. · ·

12.   H. Wang, D. W. Brandl, F. Le, P. Nordlander, and N. J. Halas, "Nanorice: a hybrid plasmonic nanostructure," Nano Letters, vol. 6, no. 4, pp. 827–832, 2006. · ·

13.   H. Qi, D. Alexson, O. Glembocki, and S. M. Prokes, "Plasmonic coupling on dielectric nanowire core-metal sheath composites," Nanotechnology, vol. 21, no. 8, Article ID 085705, 2010. · ·

14.   H. Qi, D. Alexson, O. Glembocki, and S. M. Prokes, "The effect of size and size distribution on the oxidation kinetics and plasmonics of nanoscale Ag particles," Nanotechnology, vol. 21, no. 21, Article ID 215706, 2010. · ·

15.   K. Kneipp, H. Kneipp, R. Manoharan et al., "Extremely large enhancement factors in surface-enhanced Raman scattering for molecules on colloidal gold clusters," Applied Spectroscopy, vol. 52, no. 12, pp. 1493–1497, 1998.

16.   H. Xu, J. Aizpurua, M. Käll, and P. Apell, "Electromagnetic contributions to single-molecule sensitivity in surface-enhanced Raman scattering," Physical Review E, vol. 62, no. 3, pp. 4318–4324, 2000. · ·

17.   Z. Wang, S. Pan, T. D. Krauss, H. Du, and L. J. Rothberg, "The structural basis for giant enhancement enabling single-molecule Raman scattering," Proceedings of the National Academy of Sciences of the United States of America, vol. 100, no. 15, pp. 8638–8643, 2003. · ·

18.   K. Kneipp, Y. Wang, R. R. Dasari, and M. S. Feld, "Approach to single molecule detection using surface-enhanced resonance Raman scattering SERRS: a study using rhodamine 6G on colloidal silver," Applied Spectroscopy, vol. 49, no. 6, pp. 780–784, 1995.

19. D. M. Kuncicky, S. D. Christesen, and O. D. Velev, "Role of the micro- and nanostructure in the performance of surface-enhanced Raman scattering substrates assembled from gold nanoparticles,"Applied Spectroscopy, vol. 59, no. 4, pp. 401–409, 2005.

20. P. M. Tessier, O. D. Velev, A. T. Kalambur, J. F. Rabolt, A. M. Lenhoff, and E. W. Kaler, "Assembly of gold nanostructured films templated by colloidal crystals and use in surface-enhanced Raman spectroscopy," Journal of the American Chemical Society, vol. 122, no. 39, pp. 9554–9555, 2000. · ·

21. A. Tao, F. Kim, C. Hess et al., "Langmuir-Blodgett silver nanowire monolayers for molecular sensing using surface-enhanced Raman spectroscopy," Nano Letters, vol. 3, no. 9, pp. 1229–1233, 2003. · ·

22. S. M. Prokes, O. J. Glembocki, R. W. Rendell, and M. G. Ancona, "Enhanced plasmon coupling in crossed dielectric/metal nanowire composite geometries and applications to surface-enhanced Raman spectroscopy," Applied Physics Letters, vol. 90, no. 9, Article ID 093105, 3 pages, 2007. · ·

23. COMSOL Multiphysics, COMSOL Inc., http://www.comsol.com/.

24. J. Zhao, A. O. Pinchuk, J. M. McMahon et al., "Methods for describing the electromagnetic properties of silver and gold nanoparticles," Accounts of Chemical Research, vol. 41, no. 12, pp. 1710–1720, 2008. · ·

25. M. Micic, N. Klymyshyn, Y. D. Suh, and H. P. Lu, "Finite element method simulation of the field distribution for AFM tip-enhanced surface-enhanced Raman scanning microscopy," Journal of Physical Chemistry B, vol. 107, no. 7, pp. 1574–1584, 2003. · ·

26. H. Ditlbacher, A. Hohenau, D. Wagner et al., "Silver nanowires as surface plasmon resonators," Physical Review Letters, vol. 95, no. 25, Article ID 257403, 4 pages, 2005. · ·

27. M. W. Knight and N. J. Halas, "Nanoshells to nanoeggs to nanocups: optical properties of reduced symmetry core-shell nanoparticles beyond the quasistatic limit," New Journal of Physics, vol. 10, Article ID 105006, 2008. · ·

28. M. W. Knight, Y. Wu, J. B. Lassiter, P. Nordlander, and N. J. Halas, "Substrates matter: influence of an adjacent dielectric on an individual plasmonic nanoparticle," Nano Letters, vol. 9, no. 5, pp. 2188–2192, 2009. · ·

29. T. Laroche and C. Girard, "Near-field optical properties of single plasmonic nanowires," Applied Physics Letters, vol. 89, no. 23, Article ID 233119, 3 pages, 2006. · ·

30. M. W. Knight, N. K. Grady, R. Bardhan, F. Hao, P. Nordlander, and N. J. Halas, "Nanoparticle-mediated coupling of light into a nanowire," Nano Letters, vol. 7, no. 8, pp. 2346–2350, 2007. · ·

31. C. Jiang, W. Y. Lio, and V. V. Tsukruk, "Surface enhanced raman scattering monitoring of chain alignment in freely suspended nanomembranes," Physical Review Letters, vol. 95, no. 11, Article ID 115503, 4 pages, 2005. · ·

32. J. Ding, V. I. Birss, and G. Liu, "Formation and properties of polystyrene-block-poly(2-cinnamoylethyl methacrylate) brushes studied by surface-enhanced raman scattering and transmission electron microscopy," Macromolecules, vol. 30, no. 5, pp. 1442–1448, 1997.

33. W. M. Sears, J. L. Hunt, and J. R. Stevens, "Raman scattering from polymerizing styrene. I. Vibrational mode analysis," Journal of Chemical Physics, vol. 75, no. 4, pp. 1589–1598, 1981.

34. V. Zucolotto, M. Ferreira, C. Cordeiro et al., "Unusual interactions binding iron tetrasulfonated phthalocyanine and poly(allylamine hydrochloride) in layer-by-layer films," Journal of Physical Chemistry B, vol. 107, no. 16, pp. 3733–3737, 2003. · ·

35. R. Aroca and A. Thedchanamoorthy, "Vibrational studies of molecular organization in evaporated phthalocyanine thin solid films," Chemistry of Materials, vol. 7, no. 1, pp. 69–74, 1995.

36. J. H. Kim, T. Kang, S. M. Yoo, S. Y. Lee, B. Kim, and Y. K. Choi, "A well-ordered flower-like gold nanostructure for integrated sensors via surface-enhanced Raman scattering," Nanotechnology, vol. 20, no. 23, Article ID 235302, 2009. · ·

37. C. A. Szafranski, W. Tanner, P. E. Laibinis, and R. L. Garrell, "Surface-enhanced Raman spectroscopy of aromatic thiols and disulfides on gold electrodes," Langmuir, vol. 14, no. 13, pp. 3570–3579, 1998. · ·

38. M. Takahashi, M. Fujita, and M. Ito, "SERS application to some electroorganic reactions," Surface Science, vol. 158, no. 1–3, pp. 307–313, 1985.

39. A. A. Mani, Z. D. Schultz, B. Champagne et al., "Molecule orientation in self-assembled monolayers determined by infrared-visible sum-frequency generation spectroscopy," Applied Surface Science, vol. 237, no. 1–4, pp. 444–449, 2004. · ·

40. K. T. Carron and L. G. Hurley, "Axial and azimuthal angle determination with surface-enhanced Raman spectroscopy: thiophenol on copper, silver, and gold metal surfaces," Journal of Physical Chemistry, vol. 95, no. 24, pp. 9979–9984, 1991.

41. S. Frey, V. Stadler, K. Heister et al., "Structure of thioaromatic self-assembled monolayers on gold and silver," Langmuir, vol. 17, no. 8, pp. 2408–2415, 2001. · ·

42. L. J. Wan, M. Terashima, H. Noda, and M. Osawa, "Molecular orientation and ordered structure of benzenethiol adsorbed on gold(111)," Journal of Physical Chemistry B, vol. 104, no. 15, pp. 3563–3569, 2000.

43. S. W. Han, S. J. Lee, and K. Kim, "Self-assembled monolayers of aromatic thiol and selenol on silver: comparative study of adsorptivity and stability," Langmuir, vol. 17, no. 22, pp. 6981–6987, 2001. · ·

44. A. Bilić, J. R. Reimers, and N. S. Hush, "The structure, energetics, and nature of the chemical bonding of phenylthiol adsorbed on the Au(111) surface: implications for density-functional calculations of molecular-electronic conduction," Journal of Chemical Physics, vol. 122, no. 9, Article ID 094708, 15 pages, 2005. · ·

45. C. J. Sandroff and D. R. Herschbach, "Surface-enhanced Raman study of organic sulfides adsorbed on silver: facile cleavage of S-S and C-S bonds," Journal of Physical Chemistry, vol. 86, no. 17, pp. 3277–3279, 1982.

46. L. Gunnarsson, E. J. Bjerneld, H. Xu, S. Petronis, B. Kasemo, and M. Käll, "Interparticle coupling effects in nanofabricated substrates for surface-enhanced Raman scattering," Applied Physics Letters, vol. 78, no. 6, 3 pages, 2001. · ·

47. P. Mohanty, I. Yoon, T. Kang et al., "Simple vapor-phase synthesis of single-crystalline Ag nanowires and single-nanowire surface-enhanced raman scattering," Journal of the American Chemical Society, vol. 129, no. 31, pp. 9576–9577, 2007. · ·

48. M. Moskovits, "Surface-enhanced spectroscopy," Reviews of Modern Physics, vol. 57, no. 3, pp. 783–826, 1985. · ·

49. D. S. Wang and M. Kerker, "Enhanced Raman scattering by molecules adsorbed at the surface of colloidal spheroids," Physical Review B, vol. 24, no. 4, pp. 1777–1790, 1981. · ·

50. P. J. Kottman, O. J. F. Martin, D. R. Smith, and S. Schultz, "Plasmon resonances of silver nanowires with a nonregular cross section," Physical Review B, vol. 64, no. 23, Article ID 235402, 10 pages, 2001. · ·

51. R. T. Hill, J. J. Mock, Y. Urzhumov et al., "Leveraging nanoscale plasmonic modes to achieve reproducible enhancement of light," Nano Letters, vol. 10, no. 10, pp. 4150–4154, 2010. · ·

52. J. Chen, T. Mårtensson, K. A. Dick et al., "Surface-enhanced Raman scattering of rhodamine 6G on nanowire arrays decorated with gold nanoparticles," Nanotechnology, vol. 19, no. 27, Article ID 275712, 2008. · ·

53. R. Kattumenu, C. H. Lee, L. Tian, M. E. McConney, and S. Singamaneni, "Nanorod decorated nanowires as highly efficient SERS-active hybrids," Journal of Materials Chemistry, vol. 21, no. 39, pp. 15218–15223, 2011. · ·

Optimal Size of Gold Nanoparticles for Surface-Enhanced Raman Spectroscopy under Different Conditions

Seongmin Hong and Xiao Li

Department of Chemistry, University of South Florida, Tampa, FL 33620, USA

ABSTRACT

Gold nanoparticles have been used as effective surface-enhanced Raman spectroscopy (SERS) substrates for decades. However, the origin of the enhancement and the effect of the size of nanoparticles still need clarification. Here, gold nanoparticles with different sizes from 17 to 80 nm were synthesized and characterized, and their SERS enhancement toward both 4-aminothiophenol and 4-nitrothiophenol was examined. For the same number of nanoparticles, the enhancement factor generated from the gold nanoparticles increases as the size of nanoparticles increases. Interestingly, when the concentration of gold or the total surface area of gold nanoparticles was kept the same, the optimal size of gold nanoparticles was found out to be around 50 nm when the enhancement factor reached a maximum. The same size effect was observed for both 4-aminothiophenol and 4-nitrothiophenol, which suggests that the conclusions drawn in this study might also be applicable to other adsorbates during SERS measurements.

INTRODUCTION

Surface-enhanced Raman spectroscopy (SERS) is a surface sensitive technique that provides high Raman scattering enhancement of molecules adsorbing on a rough metal surface, such as silver, gold, and copper [1,2]. Presently, there are two well-accepted theories describing the mechanism of SERS amplification: electromagnetic enhancement and chemical enhancement Electromagnetic enhancement (EME) is responsible for up to 10^6–10^7 increase of Raman scattering which occurs as the surface Plasmon gets excited by incident light and amplifies the electromagnetic field of the metal surface [1, 2]. Chemical enhancement, on the other hand, provides up to 10^2 increase of Raman scattering, and it happens when the molecule adsorbs strongly on the surface of the metal, which leads to changes of its polarizability. Due to the high enhancement, SERS technique has thus been studied extensively for the last few decades, and it provides many advantages including nondestructive nature, low detection limit, high sensitivity, and easy sample preparation. SERS has been widely applied in many different areas [3–10].

Metal nanoparticles (NPs) especially Ag and Au NPs have been widely employed in SERS because of their unique physical properties that depend on size and shape of the nanoparticles [11, 12]. Among the metals mentioned above, silver usually exhibits the highest SERS activity. It was reported that the optimal Ag nanoparticles size is 15 nm which generates the strongest SERS activity in solution [11]. Interestingly, the optimal size of Ag NPs was observed ca. 50 nm when off-resonance SERS is adopted [12]. On the other hand, gold has captured the most interests among recent studies. Gold metal is known to be biocompatible [10] and shows a strong excitation close to the IR region of light, which has attracted considerable interests in its use in biotechnological systems [13]. Various research groups have studied and reported the relationship between the enhancement and the shape and size of the immobilized gold NPs on a substrate using different analytes like 4-aminothiophenol (4-ATP) [14, 15] and 5,5'-dithiobis(2-nitrobenzioc acid) (DNBA) [16]. The results from all laboratories indicate that the SERS enhancement is highly dependent on many factors including the size of gold NPs, however, with controversial conclusions. When 4-ATP adsorbed on immobilized gold NPs, the SERS intensity from gold NPs with the size of 30 nm was lower than that from gold NPs with the size of 18 nm [14, 15]. When 4-ATP was sandwiched between Au NPs and a smooth Au substrate, the SERS intensity was found out to increase as the size of the gold NPs increases [14, 15]. When labeled gold NPs were immobilized at a gold smooth surface, 60 nm NPs result in the largest SERS enhancement [16] In addition, the SERS intensity from 4-mercaptobenzoic acid on gold nanoparticles was reported to increase as the size of the NPs increases in solution when the total number of nanoparticles was kept the same [17].

Therefore, in this study, we investigated and compared the SERS enhancement of gold NPs of different sizes using both 4-nitrothiophenol (4-NTP) and 4-aminothiophenol (4-ATP) as the target molecules to better compare our results with previous ones. The SERS intensities of target molecules were probed by adsorbing them onto gold nanoparticles of the average size of 17, 30, 40, 50, 60, and 80 nm created using further reduction method [18, 19]. SERS spectra were recorded by using 647 nm Argon-Krypton lasers on a confocal Raman microscope. Furthermore, we determined the SERS enhancement and size correlation of nanoparticles under three conditions including the same number of nanoparticles, the same total surface area of the nanoparticles, and the same concentration of gold in solution. Interestingly, different conclusions of the relationship were drawn under various conditions. The optimal size of gold NPs that provides the highest enhancement factor was found out while the concentration, and the total surface area and total number of the gold NPs were kept the same. This is the first time that the optimal size of gold NPs was established under various conditions in solution.

EXPERIMENTAL METHODS

Hydrogen Tetrachloroaurate (III) trihydrate (HAuCl$_4$·3H$_2$O, 99.9+ %), trisodium citrate (C$_6$H$_5$Na$_3$O$_7$·2H$_2$O, 99.9%), and Hydroxylamine Hydrochloride (H$_3$NO·HCl, 99+%) were purchased from Fisher Scientific and were used as received. SERS tested sample, 4-nitrothiophenol (80%), and 4-aminothiophenol (97%) were purchased from Sigma-Aldrich and were used as received. All glassware was acid washed using sulfuric acid and nitric acid, and all of the solutions were prepared using deionized water (18.2 MΩ·cm) from a Cascada BIO and AN Lab Water System.

Synthesis of Gold Nanoparticles

Gold nanoparticles (NPs) of different sizes were synthesized by multiple reduction processes based on the work of Cyrankiewicz et al. [18] and Haiss et al. [19]. First, gold nanoparticles were synthesized by reduction of 0.5 mM HAuCl$_4$ with 1% trisodium citrate in aqueous solution [18]. After this step, gold NPs of the average size of 17 nm were obtained. Next, 2.50 mL of 25.4 mM HAuCl$_4$ and 3.00 mL of 0.20 M Hydroxylamine were added into the 17 nm gold seed solution to synthesize gold NPs with a larger size. After that step, gold NPs of the average size of 30 nm were obtained. Using 30 nm gold NPs as the seeds, various amounts of 25.4 mM HAuCl$_4$ and 0.20 M Hydroxylamine were added to the solution to synthesize larger gold NPs of different sizes [19]. In detail, 0.45 mL, 0.80 mL, and 1.3 mL HAuCl$_4$ and 0.70 mL, 1.5 mL, and 2.0 mL of Hydroxylamine were used to synthesize gold NPs with the average size of 40, 50, and 60 nm, respectively. To synthesize 80 nm gold NPs, 60 nm Au NPs were used as seed and 0.43 mL HAuCl$_4$ and 0.67 mL of Hydroxylamine were added. An extra amount of reducing agent was added to ensure complete reduction of gold. All of the solutions were dark red at the end. Colloidal solutions were kept in the dark during storage because of the photosensitivity concerns [20].

Sample Preparation

In order to create calibration plots for 4-ATP and 4-NTP, samples were prepared by first pipette mixing 1.00 mL of gold colloidal solutions with different amount of 4-ATP or 4-NTP; then for 4-NTP, 1.00 mL of 1 M NaCl was added into a glass cuvette. All of the measurements were done using exactly the same experimental conditions. Prepared samples were placed in the dark for 10 minutes before the measurement. This step helps to minimize the fluctuations of SERS spectra by allowing the colloidal solution to reach the equilibrium state.

UV-Vis, SEM, and SERS Measurements

UV-Vis absorption spectra of the colloidal solution were obtained with a Beckman Coulter DU 640 spectrometer. The UV-Vis spectra were used to elucidate the relative size of the particles in the solutions by comparing the location of the maximum peak wavelength in the spectra. Gold colloidal solutions were mixed with equal parts of deionized water and placed into quartz cuvette. The concentrations of gold colloidal solutions were adjusted to offset the UV-Vis spectrum for easy viewing. The concentrations were 0.083 mM, 0.040 mM, 0.015 mM, 0.017 mM, 0.019 mM, and 0.028 mM for gold NPs of the average size from 17 nm and 80 nm. The wavenumber range was set between 250 nm to 800 nm. The resolution of 0.5 nm was used to scan the gold colloidal solutions. UV-Vis spectra of the 17 nm to 80 nm gold colloidal solutions have been collected over a 4-week period of time to test the stability of the nanoparticles.

To measure the size of gold nanoparticles, Scanning Electron Microscopy (SEM) images were obtained using the Hitachi microscope (Hitachi S-800) located at the Nanotechnology Research and Education Center on the University of South Florida's Tampa Campus. Before the SEM measurements, the gold colloidal solutions were dropped (ca. 2 µL) on top of the silicon wafer and were air dried. The wafers were kept away from light because of the photosensitivity concerns. By imaging the particles using SEM [21], the size and shape of individual particles of gold were characterized as well as the size distribution of the particles.

All the SERS experiments were carried out using a Confocal Raman Microscopy (Olympus, IX71) purchased from Horiba Jovin Yvon, equipped with an Argon and Krypton laser (Coherent, Innova 70C series) producing 514 nm and 647 nm of wavelengths. For all experiments, an excitation laser with the wavelength at 647 nm has been applied with 40 mW of power, 3 s of exposure time, and 3 accumulations. The spectrum grating was 600, and the 20X microscopic objective lens was used throughout the experiments. All of the results that were reported have been repeated independently for three times for the reproducibility. To examine the effect of nanoparticle size on the enhancement, the concentration of gold, the total number, or the total surface area of the gold NPs was kept the same in the final sample solutions.

RESULTS/DISCUSSION

SEM Images of Different-Sized Au Nanoparticles

To measure the size and its distribution of gold NPs, SEM technique was applied to get image of the gold NPs. The size of gold NPs in each image was measured individually. At least 100 nanoparticles were counted for each sample to estimate the mean diameter and the relative standard deviation of the gold nanoparticles.

Figure 1 shows typical SEM images and the histograms of size distribution of gold nanoparticles with the mean diameter of 17, 30, 40, 50, 60, and 80 nm, respectively. The particle shape was nearly spherical for the nanoparticles of all sizes. The statistical analysis results of the Au nanoparticles including the mean size, standard deviation, and relative standard deviation are listed in Table 1. The standard deviation increases as the size of the gold NPs increases. However, the relative standard deviation values for the sizes around 40, 50, 60, and 80 nm of gold NPs are apparently lower than others. The size distribution of the gold nanoparticles is comparable to previous studies [17], which allow us to correlate the size of the nanoparticles with the SERS properties of the nanoparticles in solution.

Table 1: Size distribution, maximum peak wavelength in UV-Vis absorption spectra, and calculated concentration, number, and surface area of different gold colloidal solutions

Size			$_{max}$(nm)	Concentration of gold (mM)	Number of gold NPs (/L)	Surface area of gold NPs (m²/L)
Mean (nm)	Standard deviation (nm)	Relative standard deviation (%)				
17	5	29	510	0.25	$(2.1 \pm 0.9) \times 10^{16}$	19 ± 5
30	7	23	525	0.12	$(6 \pm 2) \times 10^{15}$	15 ± 3
40	6	15	530	0.045	$(8 \pm 2) \times 10^{14}$	4.1 ± 0.6
50	8	16	536	0.050	$(4 \pm 1) \times 10^{14}$	3.6 ± 0.6
60	8	13	540	0.056	$(3.1 \pm 0.6) \times 10^{14}$	3.5 ± 0.5
80	13	16	550	0.084	$(6 \pm 3) \times 10^{13}$	1.2 ± 0.4

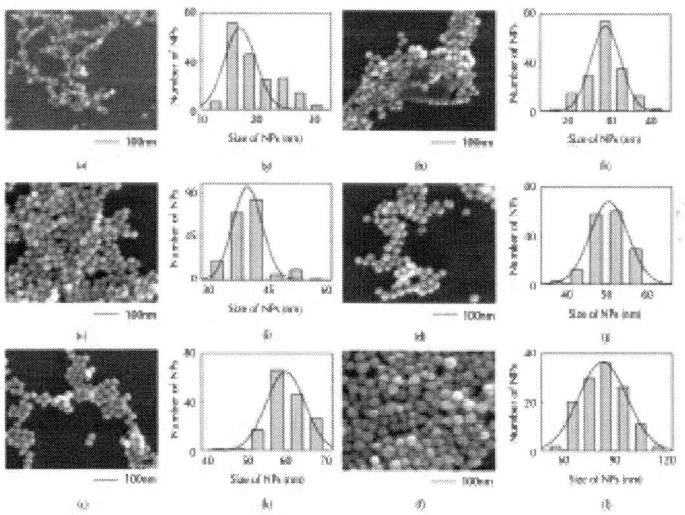

Figure 1: SEM images of the gold NPs with an average size of (a) 17 nm, (b) 30 nm, (c) 40 nm, (d) 50 nm, (e) 60 nm, and (f) 80 nm. The size histograms of the gold NPs with an average size of (g) 17 nm, (h) 30 nm, (i) 40 nm, (j) 50 nm, (k) 60 nm, and (l) 80 nm. The solid line in the size histograms is the simulation curve of Gaussian distribution.

UV-Vis Absorption Spectra of Gold Nanoparticles

Figure 2(a) shows the UV-Vis absorption spectra of gold nanoparticles of average size of 17, 30, 40, 50, 60, and 80 nm, respectively. The wavelength with maximum absorbance λ_{max} was found out to increase from 519 nm to 550 nm, as shown, in Table 1 As the size of Au NPs increases, the λ_{max} increases which agrees well with the previous conclusion that the maximum peak wavelength red-shifts as the relative particle size gets bigger [18, 22].

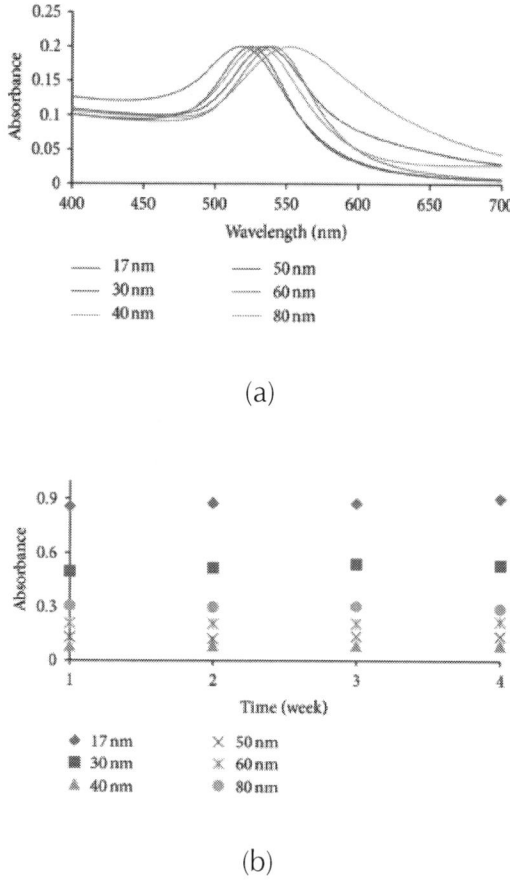

(a)

(b)

Figure 2: (a) The UV-Vis spectra of gold NPs of different size. The absorbance was normalized for better comparison. (b) The change of maximum absorbance of 17 to 80 nm gold NPs over 4 weeks. The concentrations were 0.083 mM for 17 nm, 0.040 mM for 30 nm, 0.015 mM for 40 nm, 0.017 mM for 50 nm, 0.019 mM for 60 nm, and 0.028 mM for 80 nm.

To test the stability of those gold nanoparticles, UV-Vis absorption spectra of 17 nm to 80 nm colloidal solutions were collected over a 4-week period of time. Figure 2(b) shows the change of the maximum absorbance of different gold colloidal solutions with time. Clearly, the absorption intensity of those gold colloidal solutions does not change much over a month, which indicates that gold NPs are stable within that time frame. This phenomenon is also approved by SERS measurements as the SERS activity of those gold nanoparticles changes little within the period of one month.

Once the average size and its distribution of gold NPs were determined using SEM and UV-Vis techniques, the concentration, number, and surface area of gold nanoparticles in the colloidal solutions with different sizes were calculated. Since more than enough reducing agent was added during synthesis, total reduction of gold is expected. Therefore, the number of gold NPs in the colloidal solution was estimated by (1) through dividing the total mass of gold in $HAuCl_4$ used to synthesize the gold NPs by the individual mass of a gold NP:

$$n = \frac{m_t}{m_i} = \frac{m_t}{DV} = \frac{m_t}{D4\pi r^3/3} = \frac{6m_t}{D\pi d^3}.$$

(1)

In (1), n is the number of nanoparticles; m_t is the total mass of Au in the solution; m_i is the mass of one nanoparticle; D is the density of Au assuming that the density does not change with the size of the nanoparticles; [23] r is the radius of the nanoparticle and d is the diameter of the nanoparticle.

Similarly, surface area of gold nanoparticles with certain size is calculated using (2) assuming that all nanoparticles are spherical:

$$A = \pi d^2 n = \frac{6\pi d^2 m_t}{D\pi d^3} = \frac{6m_t}{Dd}.$$

(2)

Table 1 shows the distribution of size, the calculated concentration, the number, and the surface area of the gold nanoparticles in different colloidal solutions when synthesized.

SERS Studies of 4-Aminothiophenol (4-ATP) to Test the Effect of Concentration, Number, and Surface Area of Nanoparticles on the Enhancement

SERS technique was applied to compare the enhancement of gold NPs with different sizes. Figure 3(a) shows the SERS spectra of 4-ATP collected with gold NPs with the size of 17, 30, 40, 50, 60, and 80 nm respectively. The observed peak positions of 4-ATP agree with the literature values [24, 25]. In detail, the peaks at 1079 and 1587 cm⁻¹ are assigned to the stretching vibrations of C–S and C–C, respectively. Another peak at 390 cm⁻¹ represents the bending vibration of C–S The peak at 1079 cm⁻¹, as indicated by the arrow in Figure 3(a), was used to compare the intensity of the different gold solutions because it exhibits the highest intensity and is characteristic of 4-ATP [15]. The rest of the assignment of SERS peak of 4-ATP is summarized in Table 2.

Table 2: SERS spectral peak assignment of 4-ATP[a] and 4-NTP[b]

SERS peak (cm⁻¹) of 4-ATP	Assignment	SERS peak (cm⁻¹) of 4-NTP	Assignment
390	(CS)	332	(CS)
635	g(CCC)	723	P(CH)+ P(CS)+ P(CC)
1005	g(CC)+g(CCC)	854	P(CH)
1079	n(CS)	1081	n(CS)
1177	n(CH)	1110	(CH)
1485	n(CC) + (CH)	1340	n(NO₂)

1587	n(CC)	1569	n(CC)

[a]The peak assignment of 4-ATP is from [24, 25].
[b]The peak assignment of 4-NTP is from [25].

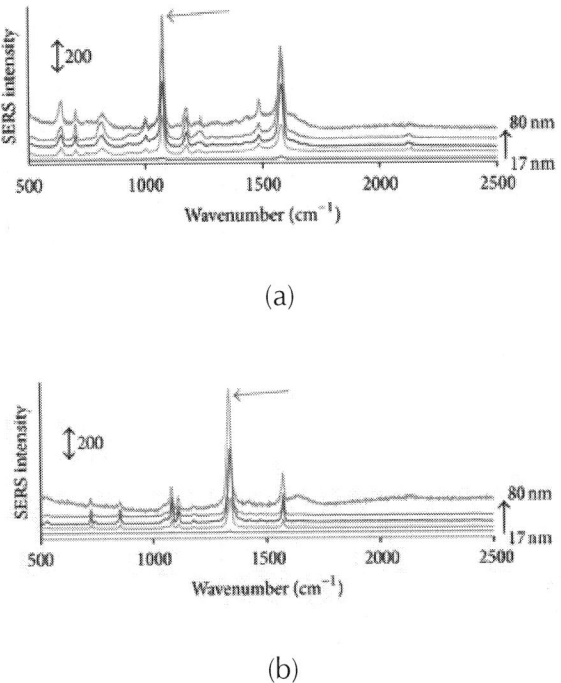

(a)

(b)

Figure 3: SERS spectra of (a) 0.25 µM 4-ATP and (b) 0.044 µM 4-NTP aqueous solution using gold nanoparticles with the size of 17, 30, 40, 50, 60, and 80 nm, respectively. The arrow of 4-ATP ($1079\,cm^{-1}$) and 4-NTP ($1340\,cm^{-1}$) indicates the peak used for the calculations of the enhancement factor.

Figures 4(a)–4(c) show the change of SERS analytical enhancement factor (AEF) of the vibration peak at 1079 cm−1 with the size of gold nanoparticles when (a) the number of gold NPs, (b) the total surface area of NPs, and (c) the concentration of gold were kept the same. The AEF shown in Figure 4 was calculated using (3), where I_{SERS} and I_{NR} are the intensity of the vibrational peak in SERS and normal Raman (NR) measurements, respectively, and C_{NR} and C_{SERS} are the concentration of 4-ATP or 4-NTP in NR measurements and the SERS measurements, respectively [26–28]:

$$EF = \frac{I_{SERS} C_{NR}}{I_{NR} C_{SERS}}.$$

(3)

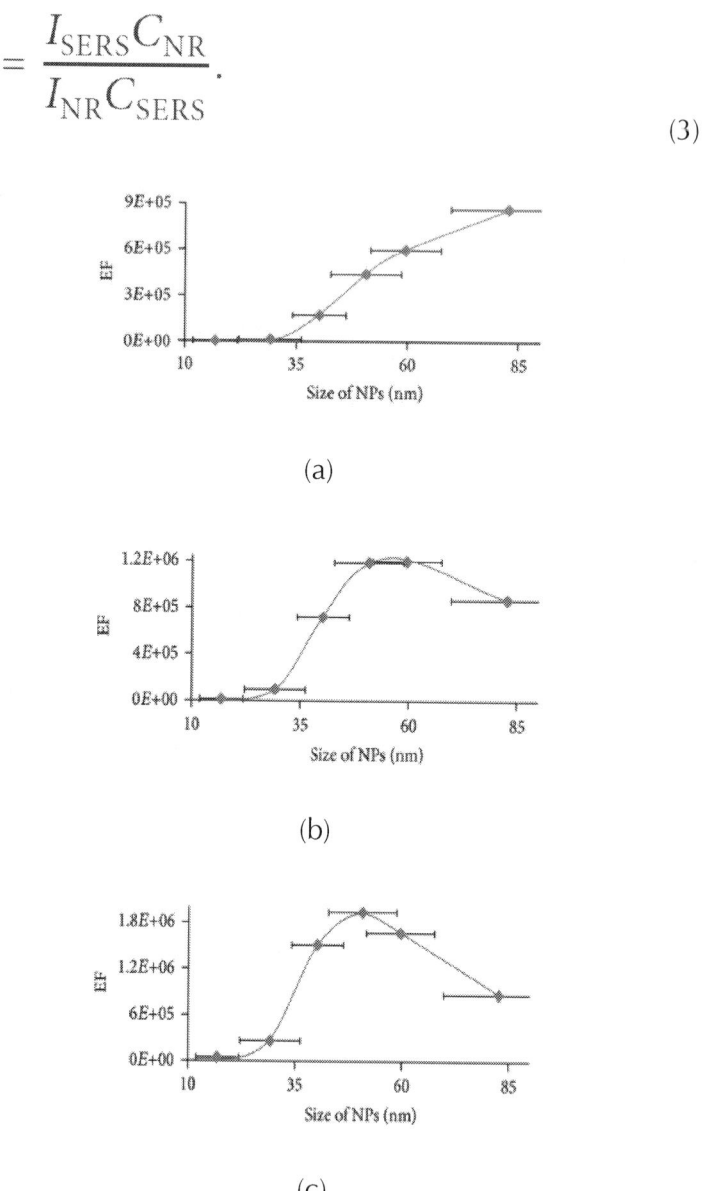

(a)

(b)

(c)

Figure 4: Experimental enhancement factor of 4-ATP, 1079 cm^{-1}, as a function of size of gold nanoparticles (a) when the number of gold NPs was kept the same, (b) when the surface area of all gold NPs was kept the same, and (c) when the concentration of gold was kept the same. The solid lines are just guiding of the eye. The error bar in x-axis is the standard deviation of the size of the nanoparticles while the error bar in y-axis is the standard deviation of enhancement factors from three experiments.

Since SERS enhancement is mainly influenced by the surface adsorbed molecules, $0.18\,\mu M$, which was located in dynamic linear range of calibration plots of 4-ATP and 4-NTP, was used for the SERS measurements for more accurate analytical enhancement factor (AEF) measurement. The dynamic linear range of the analytes for the SERS measurements is determined using calibration plots for gold NPs with different size. When the total numbers of gold NPs in the sample were kept the same, the SERS intensity increases as the size of the gold NPs increases, as shown in Figure 4(a) This observation can be easily understood due to the fact that as the size of the gold NPs increases, the surface area of the NPs increases; hence, the amount of 4-ATP molecules adsorbed on the surface increases, and so does the SERS intensity. Clearly, surface area of the SERS substrates is directly related to the enhancement achieved for a SERS measurement.

However, is the surface area the only reason for achieving high SERS enhancement? To answer this question, the spectra of 4-ATP from different colloidal solutions were collected by keeping the total surface area of the gold NPs the same. The results were shown in Figure 4(b). Interestingly, the highest enhancement was achieved from gold nanoparticles with a size ca. 50 nm. This result undoubtedly indicates that the SERS enhancement depends not only on the surface area of the SERS substrates but also on other factors such as the enhanced electromagnetic field generated from the surface plasma. It is known that the local electromagnetic enhancement increases with the increasing particle size [29]. But, as the nanoparticles size gets bigger, the convex shape, of the surface becomes flatter, the particles absorb less light and less inelastic scattering occurs on the surface, which leads to weakening of electromagnetic field on the surface and the overall SERS intensity [14, 30–32]. Moreover, previous study of the correlation between the surface plasmon resonance (SPR) properties of gold NPs and SERS spectra has revealed that the SP band red-shifts with increasing particles size. The largest SERS EF was found for the gold NPs with an SPR maximum between the wavelengths for laser excitation and for the vibrational band under study [16]. Therefore, as the size of the nanoparticles increases, the SPR moves close to the excitation wavelength of the laser (647 nm) and eventually away from that of the vibrational mode (696 nm). This explains our observations that the SERS EF generated from gold nanoparticles maximizes when the size of the gold NPs is around 50 nm. Interestingly, this conclusion agrees well with previous report when Au nanoparticles were immobilized on substrate [18].

Substrate enhancement factors (SEF) is another way of calculating enhancement factor, which is shown in (4), where I_{SERS} and I_{NR} are the intensities of the vibrational peak in SERS and normal Raman (NR) measurements, respectively, and N_{NR} and N_{SERS} are the total numbers of 4-ATP and 4- NTP in NR measurements and the surface concentration of the analytic in SERS measurements, respectively [26–28]:

$$EF = \frac{I_{SERS}N_{NR}}{I_{NR}N_{SERS}}.$$

(4)

SEF considers the actual concentration of molecules and surface area of NP substrates that contribute to the SERS enhancement. This SEF can compromise the disadvantage of AEF where you have more than a monolayer of molecules present in sample. In this experiment, SEF was calculated for the size of gold NPs that showed highest AEF. For 4-ATP, the highest AEF was found with size around 50 nm of gold for both conditions where the concentration and surface area of gold were kept the same. The SEF for gold NPs with size around 50 nm was estimated to be 2.67×10^4.

Employing gold NPs in SERS experiment opened up more opportunities to study biological samples both in vivo and in vitro because of gold's biocompatibility. In addition, gold nanoparticles have attracted more and more attention as promising drug delivery means [10, 30]. To minimize possible toxicity effect of injecting gold NPs into the biological samples, the smallest possible amount of gold should be used for each measurement. Therefore, the effect of the gold NPs' size on their SERS enhancement was tested by keeping the concentration of the gold in the sample the same. The results were shown in Figure 4(c). The highest enhancement was achieved from gold nanoparticles with a size ca. 50 nm. According to the recent study by Chen and his colleagues, gold nanoparticles with the size bigger than 8 nm and smaller than 50 nm are toxic to the biological samples that they tested [33]. Therefore, using gold nanoparticle ca. 50 nm can not only minimize the possible toxic effect on biological samples but also maximize the SERS enhancement factor generated from the NPs.

SERS Studies of 4-Nitrothiophenol (4-NTP) to Test the Effect of Concentration, Number, and Surface Area of Nanoparticles on the Enhancement

To test whether the conclusions drawn above are specific to 4-ATP and also to compare our observations with previous reports, we performed the same study using 4-NTP as the target molecule. Figure 3(b) shows the SERS spectra of 4-NTP collecting from gold NPs with the size of 17, 30, 40, 50, 60, and 80 nm, respectively [24,25]. The peaks at 1081, 1340, and 1587 cm^{-1} are due to the stretching vibrations of C–S, N–O, and C–C, respectively. The peak at 854 cm^{-1} represents the wagging vibration of C–H while the peak at 1110 cm^{-1} is assigned to the bending vibration of C–H. The peak at 1340 cm^{-1}, as indicated by the arrow in Figure 3(b), was used to compare the intensity of the different gold solution because it exhibits the highest intensity. The assignment of the SERS peaks of 4-NTP is summarized in Table 2.

Figures 5(a)–5(c) show the change of SERS AEF of the vibration peak at 1340 cm^{-1} with the size of gold nanoparticles when (a) the number of gold NPs, (b) the total surface area of NPs, and (c) the concentration of gold were kept the same. Generally, despite their similar chemical structure, big differences were observed between the SERS spectra and the EF of 4-NTP and those of 4-ATP even though the same experimental conditions including the SERS substrates were used. This was observed before and might be explained by the difference in their chemical structure between the nitro group of 4-NTP and the amine group of 4-ATP [24, 25]. As shown in Figure 5(a), when the number of gold nanoparticles was kept the same, the SERS EF increases as the size of the gold particle increases. Furthermore, Figures 5(b) and 5(c) show that the highest EF was achieved from gold nanoparticles around 50 nm when the surface area and the concentration of gold were kept the same, respectively. The SEF for 4-NTP for size 50 nm gold NPs were both 1.27 × 10^4 [26–28]. All of these results were nearly identical with those from 4-ATP. This clearly indicates that the conclusions drawn from 4-ATP and 4-NTP might also be applicable to other adsorbates when using Au nanoparticles as the SERS substrates for the detection.

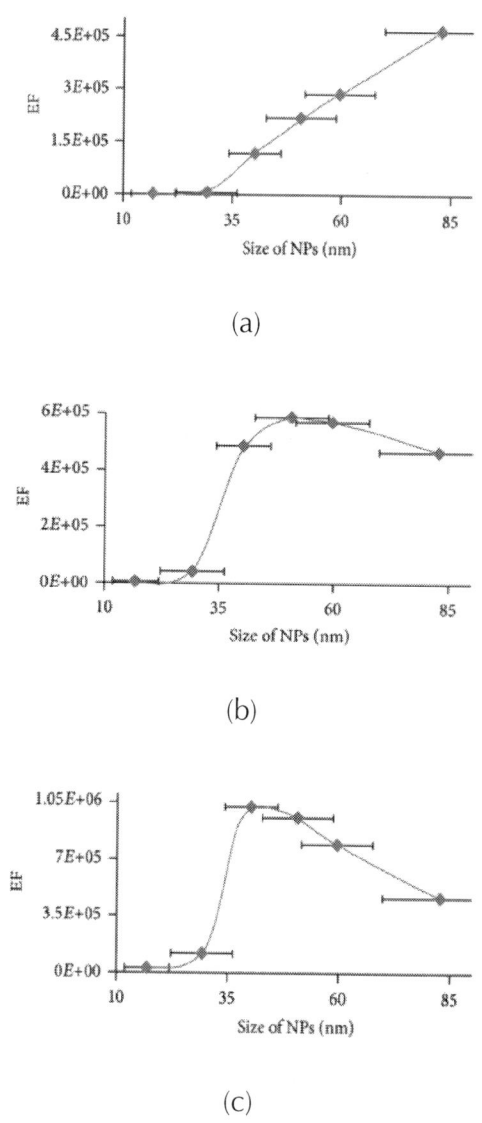

(a)

(b)

(c)

Figure 5: Experimental enhancement factor of 4-NTP, calculated from peak at $1340\,cm^{-1}$, as a function of size of gold nanoparticles (a) when the number of gold NPs was kept the same, (b) when the surface area of all gold NPs was kept the same, and (c) when the concentration of gold was kept the same. The solid lines are just guiding of the eye. The error bar in x-axis is the standard deviation of the size of the nanoparticles while the error bar in y-axis is the standard deviation of enhancement factors from three experiments.

CONCLUSIONS

Different sizes of gold nanoparticles were synthesized based on the work of Cyrankiewicz et al. [18] and Haiss et al. [19]. SEM technique was applied to determine the size and shape of the gold NPs. The average size of gold nanoparticles was found out to be 17, 30, 40, 50, 60, and 80 nm, respectively, with spherical shape. Gold NPs with the size of 50 nm exhibit the largest standard deviation. The red shift of λ_{max} value from UV-Vis absorption spectra of the colloidal solutions also indicates the size increment of gold NPs. Both the SERS activity and the UV-Vis absorption of gold NPs were observed to be stable for at least one month.

Using gold NPs with different sizes, the SERS spectra of 4-ATP and 4-NTP were collected. In order to find out the optimal size of the gold NPs that provides the highest enhancement factor, either the concentration of gold, the number, or the surface area of the gold NPs was kept the same. When the number of gold NPs was kept constantly, there was a positive linear relationship between the size of the NPs and the EF based on the SERS measurements. This phenomenon can be explained by the fact that surface area increases as the size of the gold NPs increases when the total number of NPs is the same, which leads to the higher SERS intensity. However, interesting phenomena were observed when keeping the total surface area of gold NPs or the concentration of gold constant. As shown in Figures 4(b), 4(c), 5(b), and 5(c), the highest intensity was observed when the size of gold NPs was around 50 nm when either the surface area of gold NPs or the concentration of gold was the same. Such phenomena might be explained by the relationship between the SERS enhancement and the size or the surface area of the NPs.

More importantly, the same correlations were obtained for both 4-ATP and 4-NTP. This indicates that such conclusions might not be highly sensitive to the chemical structure of the target molecules.

Instead, the conclusions might also be applicable to other adsorbates. Generally, gold NPs of the size around 50 nm are optimal when the surface area or the concentration is critical. In addition, gold NPs with size around 50 nm show minimum toxicity effect on biological samples. This conclusion is essential when SERS is used to detect important biomolecules in biological samples so that minimum amount of gold can be introduced into the biological system to achieve lowest toxicity possible while the highest SERS sensitivity can be reached at the same time.

ACKNOWLEDGMENTS

This research was supported by James & Esther King Biomedical Research Program, Florida Department of Health, and the University of South Florida starting fund. The authors thank Professor Wonkuk Kim from the Department of Mathematics and Statistics for his expertise in statistics.

REFERENCES

1. L. Xia, J. Wang, S. Tong, G. Liu, J. Li, and H. Zhang, "Design and construction of a sensitive silver substrate for surface-enhanced Raman scattering spectroscopy," Vibrational Spectroscopy, vol. 47, no. 2, pp. 124–128, 2008.

2. Y. Wang, H. Wei, B. Li et al., "SERS opens a new way in aptasensor for protein recognition with high sensitivity and selectivity," Chemical Communications, no. 48, pp. 5220–5222, 2007.

3. S. Nie and S. R. Emory, "Probing single molecules and single nanoparticles by surface-enhanced Raman scattering," Science, vol. 275, no. 5303, pp. 1102–1106, 1997.

4. R. J. Dijkstra, W. J. J. M. Scheenen, N. Dam, E. W. Roubos, and J. J. ter Meulen, "Monitoring neurotransmitter release using surface-enhanced Raman spectroscopy," Journal of Neuroscience Methods, vol. 159, no. 1, pp. 43–50, 2007.

5. J. Ni, R. J. Lipert, G. B. Dawson, and M. D. Porter, "Immunoassay readout method using extrinsic raman labels adsorbed on

immunogold colloids," Analytical Chemistry, vol. 71, no. 21, pp. 4903–4908, 1999.

6. M. Moskovits, "Spectroscopy: expanding versatility," Nature, vol. 464, no. 7287, p. 357, 2010.

7. R. A. Alvarez-Puebla and L. M. Liz-Marzán, "SERS-based diagnosis and biodetection," Small, vol. 6, no. 5, pp. 604–610, 2010.

8. I. Mannelli and M. P. Marco, "Recent advances in analytical and bioanalysis applications of noble metal nanorods," Analytical and Bioanalytical Chemistry, vol. 398, no. 6, pp. 2451–2469, 2010.

9. T. Vo-Dinh, H. N. Wang, and J. Scaffidi, "Plasmonic nanoprobes for SERS biosensing and bioimaging,"Journal of Biophotonics, vol. 3, no. 1-2, pp. 89–102, 2010.

10. J. Kneipp, H. Kneipp, B. Wittig, and K. Kneipp, "Novel optical nanosensors for probing and imaging live cells," Nanomedicine, vol. 6, no. 2, pp. 214–226, 2010.

11. C. S. Seney, B. M. Gutzman, and R. H. Goddard, "Correlation of size and surface-enhanced raman scattering activity of optical and spectroscopic properties for silver nanoparticles," Journal of Physical Chemistry C, vol. 113, no. 1, pp. 74–80, 2009.

12. K. G. Stamplecoskie, J. C. Scaiano, V. S. Tiwari, and H. Anis, "Optimal size of silver nanoparticles for surface-enhanced raman spectroscopy," Journal of Physical Chemistry C, vol. 115, no. 5, pp. 1403–1409, 2011.

13. P. K. Jain, X. Huang, I. H. El-Sayed, and M. A. El-Sayed, "Review of some interesting surface plasmon resonance-enhanced properties of noble metal nanoparticles and their applications to biosystems,"Plasmonics, vol. 2, no. 3, pp. 107–118, 2007.

14. S. C. Boca, C. Farcau, and S. Astilean, "The study of Raman enhancement efficiency as function of nanoparticle size and shape," Nuclear Instruments and Methods in Physics Research Section B, vol. 267, no. 2, pp. 406–410, 2009.

15. J. K. Yoon, K. Kim, and K. S. Shin, "Raman scattering of 4-aminobenzenethiol sandwiched between Au nanoparticles and a macroscopically smooth Au substrate: effect of size of Au

nanoparticles," Journal of Physical Chemistry C, vol. 113, no. 5, pp. 1769–1774, 2009.

16. J. D. Driskell, R. J. Lipert, and M. D. Porter, "Labeled gold nanoparticles immobilized at smooth metallic substrates: systematic investigation of surface plasmon resonance and surface-enhanced raman scattering," Journal of Physical Chemistry B, vol. 110, no. 35, pp. 17444–17451, 2006.

17. P. N. Njoki, I. I. S. Lim, D. Mott et al., "Size correlation of optical and spectroscopic properties for gold nanoparticles," Journal of Physical Chemistry C, vol. 111, no. 40, pp. 14664–14669, 2007.

18. M. Cyrankiewicz, T. Wybranowski, and S. Kruszewski, "Study of SERS efficiency of metallic colloidal systems," Journal of Physics: Conference Series, vol. 79, no. 1, Article ID 012013, 2007.

19. W. Haiss, N. T. K. Thanh, J. Aveyard, and D. G. Fernig, "Determination of size and concentration of gold nanoparticles from UV-Vis spectra," Analytical Chemistry, vol. 79, no. 11, pp. 4215–4221, 2007.

20. E. Y. Hleb and D. O. Lapotko, "Photothermal properties of gold nanoparticles under exposure to high optical energies," Nanotechnology, vol. 19, no. 35, Article ID 355702, 2008.

21. J. E. Herrera and N. Sakulchaicharoen, "Microscopic and spectroscopic characterization of nanoparticles," Drugs and the Pharmaceutical Sciences, vol. 191, p. 239, 2009.

22. J. A. Creighton, "Metal colloids," in Surface Enhanced Raman Scattering, R. K. Chang and T. E. Furtak, Eds., p. 315, Plenum, New York, NY, USA, 1982.

23. Z. Kolska, J. Riha, V. Hnatowicz, and V. Svorcik, "Lattice parameter and expected density of Au nano-structures sputtered on glass," Materials Letters, vol. 64, no. 10, pp. 1160–1162, 2010.

24. K. Kim and H. S. Lee, "Effect of Ag and Au nanoparticles on the SERS of 4-aminobenzenethiol assembled on powdered copper," Journal of Physical Chemistry B, vol. 109, no. 40, pp. 18929–18934, 2005.

25. M. E. Abdelsalam, "Surface enhanced raman scattering of aromatic thiols adsorbed on nanostructured gold surfaces," Central European Journal of Chemistry, vol. 7, no. 3, pp. 446–453, 2009.

26. V. S. Tiwari, T. Oleg, G. K. Darbha, W. Hardy, J. P. Singh, and P. C. Ray, "Non-resonance SERS effects of silver colloids with different shapes," Chemical Physics Letters, vol. 446, no. 1–3, pp. 77–82, 2007.

27. E. C. Le Ru, E. Blackie, M. Meyer, and P. G. Etchegoin, "Surface enhanced Raman scattering enhancement factors: a comprehensive study," Journal of Physical Chemistry C, vol. 111, no. 37, pp. 13794–13803, 2007.

28. W. B. Cai, B. Ren, X. Q. Li et al., "Investigation of surface-enhanced Raman scattering from platinum electrodes using a confocal Raman microscope: dependence of surface roughening pretreatment,"Surface Science, vol. 406, no. 1–3, pp. 9–22, 1998.

29. K. L. Kelly, E. Coronado, L. L. Zhao, and G. C. Schatz, "The optical properties of metal nanoparticles: the influence of size, shape, and dielectric environment," Journal of Physical Chemistry B, vol. 107, no. 3, pp. 668–677, 2003.

30. S. Ghoshal, D. Mitra, S. Roy, and D. D. Majumder, "Biosensors and biochips for nanomedical applications: a review," Sensors and Transducers Journal, vol. 113, no. 2, pp. 1–17, 2010.

31. S. A. Maie, Plasmonics: Fundamentals and Applications, Springer, New York, NY, USA, 1st edition, 2006.

32. M. Moskovits, "Surface-enhanced spectroscopy," Reviews of Modern Physics, vol. 57, no. 3, pp. 783–826, 1985.

33. Y. S. Chen, Y. C. Hung, I. Liau, and G. S. Huang, "Assessment of the in vivo toxicity of gold nanoparticles," Nanoscale Research Letters, vol. 4, no. 8, pp. 858–864, 2009.

Molecular Dynamics Analysis of PVA-AgnP Composites by Dielectric Spectroscopy

J. Betzabe González-Campos[1], Evgen Prokhorov[2],
Isaac C. Sanchez[3], J. Gabriel Luna-Bárcenas[2],
Alejandro Manzano-Ramírez[2], Jesús González-
Hernández[4], Yliana López-Castro[1],
and Rosa E. del Río[1]

[1]Department of Chemistry, Institute of Chemical and Biological Researches, Universidad Michoacana de San Nicolás de Hidalgo, Ciudad Universitaria, 58060 Morelia, MICH, Mexico

[2]Biomaterials Laboratory, Centro de Investigación y de Estudios avanzados del IPN, Unidad Querétaro, 76230 Querétaro, QRO, Mexico

[3]Department of Chemical Engineering, the University of Texas at Austin, Austin, TX 78712, USA

[4]Nanostructurated Materials Laboratory, Centro de Investigación en Materiales Avanzados, S.C., 31109 Chihuahua, CHIH, Mexico

ABSTRACT

The molecular dynamics of PVA/AgnP composites were studied by dielectric spectroscopy (DS) in the 20–300°C temperature range. Improper water elimination leads to misinterpretation of thermal relaxations in PVA composites in agreement with the previous report for pristine PVA. The evaporation of water and its plasticizing effect are more evident in pure PVA confirming the existence of strong interaction between OH groups of PVA chains and AgnP. Dry films show a single nonlinear VFT dependence (from 45°C until melting) associated to the α-relaxation and, therefore, to the glass transition phenomenon and from dielectric measurements, the T_g of composites vary from 88°C for pristine PVA to 125°C for PVA/AgnP (5 wt%). Below 45°C, dry films exhibit a single Arrhenius behavior showing a 3D hopping conductivity as explained based on the variable range hopping model. PVA/AgnP composites have higher conductivity compared to pristine PVA, and it increases as AgnP weight percent increases. Finally, DMA measurements support the statement that a secondary relaxation was erroneously assigned as the glass transition of PVA and composites in previous reports.

INTRODUCTION

Nowadays metal-polymer Nano composites are the subject of increased interest because they combine the features of polymers with those of metals. Metallic nanoparticles incorporated in or with polymers have attracted much attention due to their distinct optical, electrical, and catalytic properties, which have potential applications in different fields such as bioengineering, photonics, and electronics [1–5]. Among different metals used for nanoparticles preparation, silver is very attractive because it exhibits the highest electrical and thermal conductivities, together with their antibacterial activity and even their interaction with HIV-1 virus [6]. On the other hand, polyvinyl alcohol (PVA) has been widely used as a matrix for preparation of Nano composites due to its easy process ability and high optical clarity. It is considered among the best polymers as host matrix for silver nanoparticles (AgnP), and it is frequently used as a stabilizer due to its optical clarity, which enables investigation of the nanoparticle formation [7, 8]. PVA is a biologically friendly polymer since it is water soluble and has extremely low cytotoxicity, which allows the application of PVA-based composites in the biomedical field.

PVA/AgnP is a very attractive combination since these composites have high mechanical strength, water-solubility, good environmental stability, easy processability, and electrical conductivity [9–11]. Different studies have been carried out about the optimal parameters for the synthesis of nanoparticles, the antibacterial activity of composites, their mechanical properties, and the chemical interaction between PVA and AgnP [12–15]. However, the molecular dynamics analysis by dielectric relaxation studies and the electrical conductivity behavior of PVA-Ag nanocomposite films have been scarcely studied; there is only one report in this regard [16].

One parameter that can be characterized by means of the molecular dynamics analysis is the glass transition temperature (T_g). In polymers, polymers blends, and composites, an accurate characterization of the T_g plays a crucial role, since it indicates the change from the glassy state into a liquid or a rubbery state, and it can be a measure of compatibility or miscibility in polymer blends [17]. Additionally, physicochemical properties of a material such as dissolution, bioavailability, processing, and handling qualities can be related to the glass transition temperature of the material [18]. Also the optimal application and processing temperatures for polymers base materials are dependent on their glass transition temperature.

On the other hand, in a previous work [19], it was shown by dielectric and dynamic mechanical analysis that the nature of pure PVA thermal relaxations was erroneously assigned, since improper water elimination and the narrow temperature range analyzed in all the previous reports have led to misinterpretation of thermal relaxations in pure PVA. Commonly, the molecular dynamics analysis of PVA composites, and its blends with other polymers and inorganic compounds has been carried out in a narrow temperature range and based on pure PVA thermal relaxations behavior; however, if the molecular dynamics of pure PVA has been misunderstood, it could also be the case of the molecular dynamics of its composites and blends.

Based on these arguments, the aim of this work is twofold: to study the molecular dynamics of PVA/AgnP composites, which have been scarcely reported, and to make a comparison based on the previous studies on pure PVA to probe that previous reports on PVA-inorganic materials composites have also been misunderstanding because they were based in pure PVA results.

EXPERIMENTAL METHODS

Films Preparation

Poly (vinyl alcohol), Mw 89,000–98,000 g/mol and hydrolysis degree >99%, was purchased from Sigma-Aldrich and used as received. Carbon-coated silver nanoparticles powder 25 nm average particle size was purchased from nanotechnologies, Inc. PVA films were obtained by dissolving a known amount of PVA in water to obtain a 7.8 wt % solution under stirring at 90°C. The proper quantity of silver nanoparticles powder (0.5, 1, 2, 3 and 5% w/w respect to PVA dry based) was poured into the 7.8 wt % aqueous PVA solution, this solution was mechanically stirrer for 40 min and further sonicated during 30 min to obtain a homogeneous nanoparticle solution. Afterward PVA/AgnP films were prepared by the solvent casting method, by pouring the solutions into plastic Petri dishes and allowing the solvent to evaporate at 37°C during 24 hours. These films had thicknesses of ca. 40 μm measured by a Mitutoyo micrometer. A thin layer of gold was vacuum-deposited onto both film sides to serve as electrodes. Rectangular small pieces (about 4 mm × 3 mm) of these films were prepared for measurements, and the contact areas were measured with a digital calibrator (Mitutoyo).

Infrared Measurements and Morphology Analysis

Chemical analysis of PVA/AgnP composites was performed by FTIR on a Perkin-Elmer spectrophotometer using an ATR accessory in the range 4000–650 cm^{-1}. Resolution was set to 4 cm^{-1}, and the spectra are an average of 32 scans. PVA/AgnP films morphology was analyzed by JEOM JSM-7401F field emission scanning electron microscope.

Thermal Measurements

Moisture content was determined by thermogravimetric analysis (TGA). The moisture content was evaluated by the decrease of sample weight during the heating scan. TGA curves were obtained using a Mettler Toledo apparatus, model TGA/SDTA 851e, with a sample mass of ca. 3 mg and an aluminum sample holder under argon atmosphere with a flow rate of 75 mL/min. Heating rate was set to 10°C/min.

Dielectric Measurements

The dielectric measurements in the frequency range from 0.1 Hz to 1 MHz were carried out using a Solartron 1260 impedance gain-phase analyzer with 1294 Impedance interface and in the frequency range 100 Hz–110 MHz using an Agilent Precision Impedance Analyzer 4294A. The amplitude of the measuring signal was 100 mV. The home-made impedance vacuum cell was used in conjunction with a Watlow's Series 982 microprocessor with ramping temperature controller for all dielectric measurements from 20°C to 300°C, and for some samples with an additional thermal treatment at 120°C to obtain "dry" samples. Each sample was left at each temperature for 3 min to ensure thermal equilibrium.

Dynamic Mechanical Analysis (DMA)

DMA measurements were carried out using an RSAIII, TA Instruments with a heating rate of 5°C/min at a frequency of 0.1 Hz, under dry air atmosphere in the 25–300°C temperature range.

RESULTS AND DISCUSSION

Morphological Analysis

The AgnP dispersion on PVA matrix analyzed by SEM is shown in Figure 1. At lower concentration, a homogeneous dispersion of AgnP is observed since they are well distributed within the PVA matrix with particles size of 25 nm. When AgnP concentration increases, the formation of agglomerates cannot be avoided; however, the dispersion of silver nanoparticles is adequate and clusters are not significant. Uniform metal nanoparticles dispersion is required to guarantee the homogeneity on properties. The used methodology gives well dispersed films even at high concentration such as 5% wt.

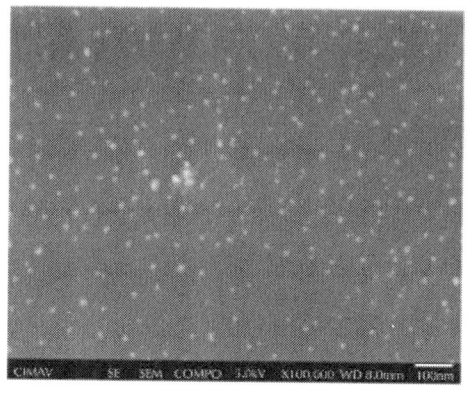

(a)

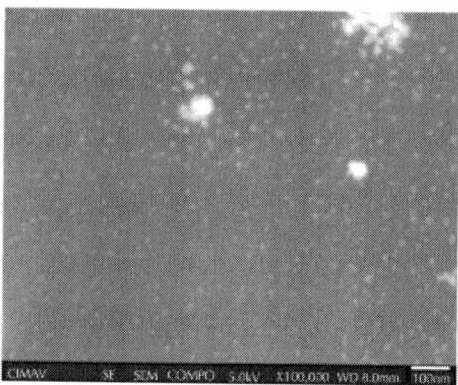

(b)

Figure 1: (a) PVA/AgnP (2%) and (b) PVA/AgnP (5%) films.

It could be seen that PVA is an excellent host matrix for encapsulation of silver nanoparticles acting as a good capping agent and providing environmental and chemical stability.

FTIR-ATR Analysis

The FTIR spectra for pristine PVA and PV/AgnP composites are shown in Figure 2. The spectrum of pristine PVA shows its characteristic bands at 1040 cm^{-1} (C–O) in acetyl group, 1090 cm^{-1} (C–O), 1140 cm^{-1} (C–O, crystallinity), 1170 cm^{-1} (C–O–C) in ether group, 1245 cm^{-1} (C–O–C) in acetyl group, 1320 cm^{-1} (CH + OH interaction), the 1375 cm^{-1} band due to the coupling of O–H vibrations at 1420 cm^{-1} with the wagging vibrations (CH$_2$), 1650 cm^{-1} (C=C), 1730 cm^{-1} (C=O), 2850 cm^{-1} (CH), 2900–2950 cm^{-1} (CH$_2$), and 3000–3500 cm^{-1} (O–H).

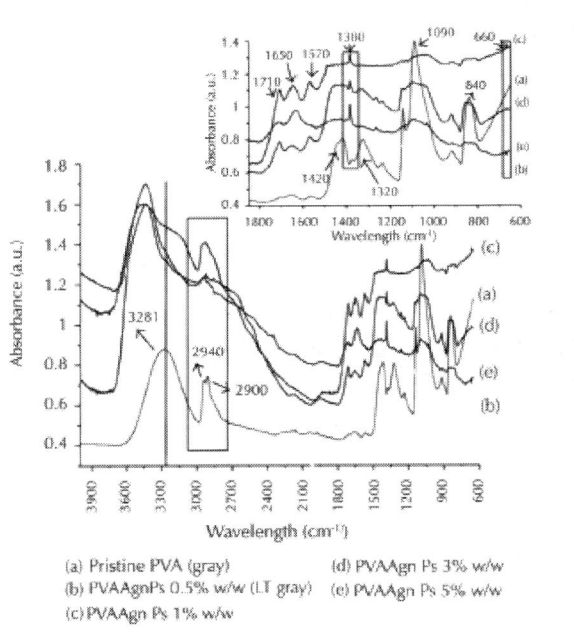

(a) Pristine PVA (gray)   (d) PVAAgn Ps 3% w/w
(b) PVAAgnPs 0.5% w/w (LT gray)   (e) PVAAgn Ps 5% w/w
(c) PVAAgn Ps 1% w/w

Figure 2: IR spectra of pristine PVA and PVA/AgnP composites. Window insert: zoom for the 1800 cm^{-1} to 600 cm^{-1} region.

Figure 2 also shows clear evidence of chemical interactions bonding between PVA and silver nanoparticles suggesting that these interactions are mainly between –OH groups of PVA and AgnP. This is observed by a general change in the intensity of the absorption bands in the stretching vibrations of OH groups involved in hydrogen bonding (3800–3000 cm^{-1}) and their shifting to 3400 cm^{-1}. The PVA-AgnP's spectrums show a new band at 660 cm^{-1} corroborating the formation of hydrogen bonds in all structures. A change in the intensity of the band at 1380 cm^{-1}, compared with the band at 1420 cm^{-1}, indicates the decoupling between OH and CH vibrations due to bonding interaction between OH and silver nanoparticles.

Moisture Content

For pristine PVA as well as for PVA/AgnP composites, two different moisture contents according to different sample treatment were evaluated: wet samples (without annealing treatment) and annealed samples at 120°C. Samples annealed at 120°C were obtained by the following fashion: a first scan was performed from 20°C to 120°C holding them at 120°C during 30 minutes and followed by cooling at 20°C, afterward a second heating from 20°C to 250°C was carried out in the same sample. Before annealing, a single scan on wet samples was performed for comparison. In the case of dry films, water content is reduced to the minimum possible by the heat treatment at 120°C during 30 minutes under controlled atmosphere inside the measuring cell.

Thermo gravimetric measurements for pristine PVA are shown in Figure 3. The curve labeled as "wet PVA" corresponds to the first scan described above (from 20 to 120°C), while that labeled as "dry PVA" is the second scan of the same film after heat treatment at 120°C during 30 minutes. The water loss is about 4.16% and 0.01% for wet and dry samples, respectively; these results show that an important quantity of water is present before any measurement of PVA films. It is noteworthy that water content in PVA/AgnP composites is very close to that reported for pristine PVA; however, it slightly decreases as AgnP increases (from 4.0% for 0.05% of AgnP to 3.6% for 5% of AgnP). For all samples studied, the second scan after annealing at 120°C (dry samples) reveals water loss about 0.01%; therefore, samples annealed at 120°C do not need further heating treatment since at this water content, it is considered that samples are in the dry state.

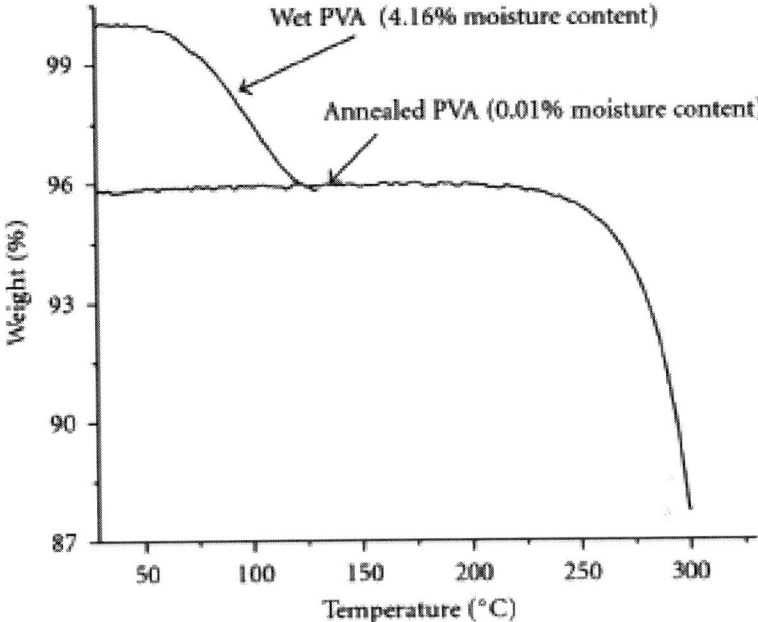

Figure 3: Thermogravimetric measurements in pristine PVA.

These same conditions labeled as wet and dry samples were applied in dielectric and DMA measurements discussed below.

Conductivity and Dielectric Results

The dc conductivity (σ_{dc}) of PVA and PVA/AgnP films was calculated by dielectric measurements using the methodology previously described elsewhere [20–22]. The dc resistance (R_{dc}) was obtained from the intersection of the semicircle and the real-part axis on the impedance plane (at $Z'' = 0$) as it is pointed out in the window insert of Figure 4, hence σ_{dc} can be calculated by the following relationship: $\sigma_{dc} = d/(R_{dc} \times A)$, where d are the thickness and A the area of the film, respectively. The Cole-Cole plot (real (Z') versus imaginary ($7''$) parts of the impedance) shows the high frequency region semicircle, which is related to the bulk effect of PVA composites, while the linear region in the low frequency range is attributed to contacts effect [20].

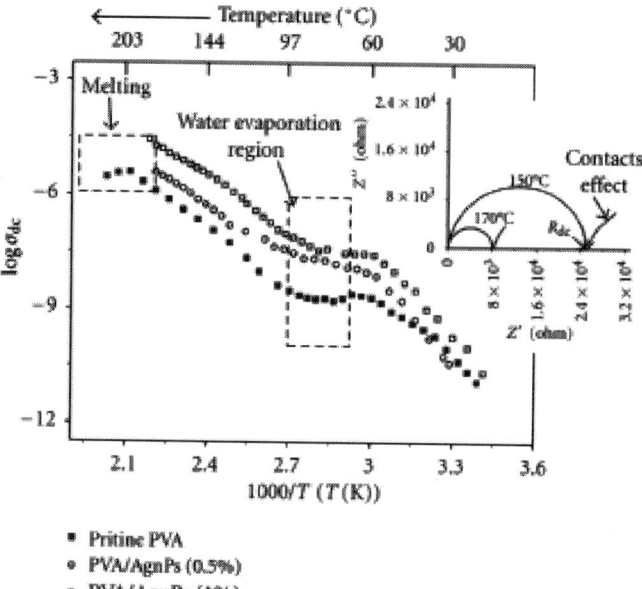

Figure 4: Conductivity (σ_{dc}) versus 1000/T(K) for pure PVA and PVA/AgnP wet composites.

As showed by TGA analysis, at ambient conditions water content in films is ca. 4.16 wt%, which reduces film resistance and masks the real conductivity behavior of composites. This is a delicate issue that needs to be considered when performing dielectric measurements especially in PVA films [19]. Due to this observation, dielectric measurements were carried out on dry films annealed in vacuum cell.

Figure 4 shows the change in dc conductivity as a function of temperature from 25 to 250°C for PVA and PVA/AgnP (0.5 and 1%) wet films. It can be seen in both films that conductivity increases as temperature increases due to the increased free volume and their respective ionic and segmental mobility [23]. This dependence unveils two well-defined regions at "low" (from 20°C to 80°C) and high (from 100°C to melting) temperatures, with an intermediate discontinuity between 80°C and 100°C associated to moisture evaporation [19]. Both relaxation regions disclose a well-defined non-Arrhenius behavior usually observed in many glass-formers and well described by the well-known Vogel-Fulcher-Tammann (VFT) relationship:

$$\sigma_{dc} = \sigma_0 e^{[-D/(T-T_0)]},$$

(1)

$\sigma_{dc} = \sigma_0 [-D/(T-T0)]$,(1)where σ_{dc}, σ_0, and D are the conductivity, the preexponential factor, and a material constant, respectively, and T_0 is the so-called Vogel temperature related to the glass transition temperature [20]. This VFT-behavior is a clear evidence for the glass transition phenomenon; however, at this point this trend is not disclosed in the whole temperature range as in many amorphous polymers such as polypeptides [24]. At this point, it would be ambiguous to assign a T_g value of plasticized PVA and composites even though the plasticization of T_g is evident, the evaporation of water could be interfering.

Previously, these two regions were erroneously described as two Arrhenius-type relaxations in pristine PVA and Gd doped-PVA [25]; however, in those composites, moisture was not properly eliminated and the analysis was performed in a narrow temperature range (from 30°C to 180°C for composites and from 30°C to 160°C for pristine PVA), this do not allow to have a wider panorama of PVA and Gd doped-PVA relaxations giving rise to the erroneous interpretation. It is also important to mention that Hanafy did not discuss the evaporation of water zone (indicated in Figure 4), even when it is clearly observed in the ac conductivity versus the reciprocal temperature plot for PVA and Gd doped-PVA (see [25]).

From Figure 4, it is noteworthy that the evaporation of water (about 80°C–100°C) is more evident in pure PVA. This event confirms the existence of strong interactions between water and pristine PVA chains, and it clearly indicates the existence of strong hydrophilic groups acting as primary hydration sites: OH side groups. In pristine PVA, an overall increase in the molecular mobility with increasing water content occurs. This water evaporation region arises as a consequence of loosely bound water molecules connected to the reorientation of water molecules in water clusters around the primary hydration sites. The hydroxyl groups exert strong effects on the PVA molecular dynamics since the interactions between OH neighbors and absorbed moisture. The evaporation of water is less evident in PVA/AgnP composites; the conductivity slightly changes in the 80°C–100°C region, but it also represents a transition region from the low temperature relaxation to the high temperature relaxation.

As it was pointed out by IR analysis, the Ag nanoparticles directly interact with OH groups in PVA chains. This interaction decreases the number of PVA-water hydrogen bonds, reducing the number of available OH primary hydration sites to attach water molecules and consequently decreasing the moisture content. Since less amount of water is present in PVA/AgnP composites, the evaporation of water region is less evident. An additionally VFT-low temperature relaxation (between 20°C and 80°C) in all PVA/AgnP wet samples shows lesser curvature, thus indicating that the plasticizing effect of water is less effective.

On the other hand, wet PVA/AgnP composites have higher conductivity compared with pristine PVA, and it increases as AgnP weight percent increases as it can be seen in Figure 4. The formation of charge transfer complexes (CTCs) by the inclusion of metal nanoparticles cause reduction of the crystalline-amorphous interface decreasing the interfacial barrier and increasing the transition probability of electron hopping across the barrier and insulator chains, which in turn provides a conducting path through the amorphous regions of the polymer matrix resulting in enhanced conductivity in agreement with Mahendia et al. [16].

At the same time, the melting temperature of PVA is affected by the inclusion of AgnP; in PVA/AgnP composites, the melting temperature shifts to lower values. This can be explained as follows: the crystallinity of PVA results from strong intermolecular and intermolecular hydrogen bonding between PVA chains mainly through –OH groups. As it was shown by FTIR measurements, AgnP interact with –OHgroups of PVA; therefore, these AgnP-OH interactions decrease the intermolecular hydrogen bonding in PVA chains affecting its crystallinity, which in turns directly affects the melting temperature. This observation has been previously studied by our group [19].

On the other hand, based on previous studies on pure PVA [19], dry films of PVA/AgnP composites were obtained; in this case, water content is reduced to the minimum possible by a thermal treatment at 120°C during 30 minutes under controlled atmosphere inside the measuring cell (water content ca. 0.01%, determined by TGA analysis). After this thermal treatment, the films are cooled down at room temperature without opening the measuring cell, and a second heating is carried out on the same films. Now the conductivity behavior of PVA/AgnP composites is similar to the results previously reported on pristine dry PVA [19], disclosing a different trend compared with wet films, as it can be seen in Figure 5.

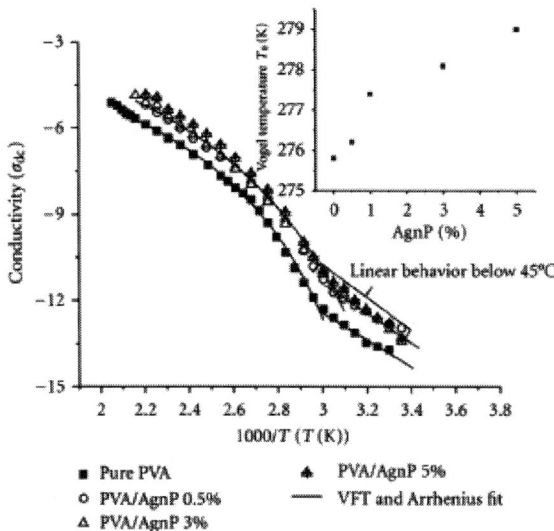

■ Pure PVA ▲ PVA/AgnP 5%
○ PVA/AgnP 0.5% — VFT and Arrhenius fit
△ PVA/AgnP 3%

Figure 5: DC conductivity versus 1000/T(K) for dry PVA/AgnP films (water content near zero). Window insert: Vogel Temperature versus AgnP % wt.

The second conductivity scan of dry films shows a single nonlinear VFT dependence from 45°C until melting, preceded by a linear behavior associated to a secondary relaxation process. This linear behavior will be discussed later.

The nonlinear behavior clearly described the glass transition phenomenon in PVA and PVA composites, thus, a single Vogel temperature can be now calculated from the fitting of the experimental data to the VFT model. This results support the statement about the plasticizing effect of water in the 20°C–80°C region and the evaporating of water region in the 80°C–100°C temperature range previously described.

Figure 5 reveals that when PVA/AgnP composites are analyzed in a broad temperature range, it is possible to observe a wider panorama on the true nature of the relaxation processes in PVA and PVA composites. In general, all reported studies on PVA composites and blends were reported up to 180°C.

As mentioned before, there is only one report on PVA-AgnP dielectric measurements [16]; however, it presents conductivity and dielectric studies as a function of AgnP concentration with no temperature variation; there are no previous reports on conductivity and dielectric measurements temperature dependencies regarding PVA-AgnP composites. Nonetheless, several dielectric studies were performed on PVA and PVA composites; however, most of them were carried out below 110°C [16, 23, 25–29]. In these studies, both VFT and Arrhenius behaviors have been reported in pure PVA and its composites in the same narrow temperature range.

Agrawal and Awadhi [26] and Linares et al. [27] showed the VFT behavior for the temperature dependence of conductivity in pure PVA and PVA-based gel electrolytes in the 20–100°C temperature range, and in pure PVA and in PVDF/PVA blends in the 20–110°C temperature range respectively. On the other hand, Hanafy [25] reported two linear behaviors in the 20°C to 180°C temperature range for PVA and Gd-doped PVA. Bhargav et al. [23] suggest an Arrhenius behavior in pure PVA and PVA:NaI complex in the 20–100°C temperature range, as well as Hema et al. [28] in the 20–70°C temperature range in pure PVA and PVA-NH4Br complexes. Zhang et al. [29] showed the relaxation time temperature dependence of PVA/MWCNT composites with 3 wt% MWCNT suggesting an Arrhenius-type dependence in the 20°C to 145°C temperature range. In the latter case, no results for pure PVA are reported. Finally, Mahanadi et al. [16] carried out electrical conductivity and dielectric spectroscopy studies of PVA-Ag Nano composite films; they observed a strong influence of the concentration of silver nanoparticles on the electrical conductivity and dielectric properties of PVA matrix. However, these results are at room temperature and they do not show temperature dependencies of conductivity and dielectric results. Despite of all the above studies, one cannot conclude the existence of a non-Arrhenius, α-relaxation process.

The nature of the molecular dynamics of Arrhenius-type and VFT-type behaviors is very different; it is very important to establish which of them is truly present in a polymer system. Arrhenius-type relaxation is related to the ions that are decoupled from the polymer host, and activated hopping is required for ionic transport. VFT behavior describes the cooperative motion, which occurs when the system is in the vicinity of a glass-transition.

In the case of PVA and PVA composites, VFT behavior was not observed in a number of previous studies on PVA composites because the effect of water was not properly taken into account [23, 25, 28, 29]. The relaxation analysis on those studies was based on temperature ranges not exceeding 160°C; most of them were done up to 100°C. These two factors, moisture and temperature range, are extremely important to gain a better understanding of PVA composites molecular dynamics.

In this work, it is clearly observed that a wider temperature range (in this case from 20°C to 250°C) and an adequate moisture content elimination allow to disclose the true nature of the molecular dynamics in PVA and PVA composites; in this case, the VFT behavior related to the glass transition.

The Vogel temperature (T_0) calculated from de VFT relationship (1) as a function of AgnP content for PVA/AgnP composites is plotted in the window insert of Figure 5 and it is shown in Table 1. In general, it can be observed that T_0 increases as AgnP content increases. For most polymers, the relationship between the glass transition temperature T_g and T_0 is $T_0 = T_g + C$, where C is an empirical constant. In many dielectric studies, C is taken as 70 K [30–32]. Table 1 shows the estimated T_g values for pristine PVA and PVA/AgnP composite as a function of AgnP content (C value is taken as 70 K).

Table 1: Vogel and glass transition temperatures for different AgnP content

Sample	T_0 (K)	T_g (°C)
Pristine PVA	275.8	72.65
PVA / AgnP (0.5 wt%)	276.8	73.15
PVA/AgnP (1 wt%)	277.5	74.35
PVA/AgnP (3 wt%)	278.1	74.95
PVA/AgnP (5 wt%)	279.0	75.85

Therefore, we can say that the glass transition temperature of PVA composites increases as AgnP weight percent increases in contrast with previous results in Ag-PVA films [33]. Moisture content of PVA composites is lower than that of pristine PVA due to the interaction between OH groups and AgnP; higher glass transition temperature values for PVA composites are expected. AgnP bonds to hydroxyl groups reduce the plasticizer effect and complicate molecular motion in PVA chains resulting in an increase in T_g values.

On the other hand, it has recently been recognized that contact and interfacial polarizations are to account when analyzing dielectric spectra. Indeed, it is well known that dc conductivity strongly affects the loss factor ε'' in the low-frequency range, and a correction must be applied to unmask the polymer dielectric effects [20]. dc Conductivity and contact polarization effect could mask the real dielectric relaxation processes in the low frequency range; therefore, to analyze the dielectric processes in detail, the complex permittivity $\varepsilon*$ is converted to the complex dielectric modulus $M*$ by the following equation: $M*=1/\varepsilon*=M'+iM''=[\varepsilon'/(\varepsilon'^2+\varepsilon''^2)+i\varepsilon''/(\varepsilon'^2+\varepsilon''^2)]$, where M' and M'' are the real and imaginary parts of electric modulus and ε' and ε'' are the real and imaginary part of permittivity, respectively. The dielectric modulus is commonly used to analyze dielectric experimental data because interfacial polarization, electrode contribution, and conductivity dc effect do not affect M'' peak. Note that M'' is temperature dependent [34].

The relaxation time dependence on temperature shown in Figure 6 was obtained from the maximum of the imaginary part of the dielectric modulus ($M*$), with the M'' peak ($\tau=1/2\pi f_{max}$) calculated at each temperature. The wet films display the trend akin to conductivity as temperature increases: two VFT behaviors in the high and low temperature range separated by moisture evaporation effect in the 80°C to 100°C temperature range. The T_0 calculated from the VFT model for relaxation time: $\tau_{dc}=\tau_0^{[D/(T-T0)]}$ (where τ, τ_0, and D are the relaxation time, the preexponential factor, and a material constant, respectively, and T_0 is the so called Vogel temperature) for all composites are shown in window insert of Figure 6. In dry films, once again the Vogel temperature increases as AgnP concentration increases and relaxation times increase as AgnP concentration increases, and these values are in agreement with those calculated above by the conductivity plot. The interaction of AgnP with OH groups hinders the mobility of PVA chains resulting in higher relaxation times for this primary relaxation.

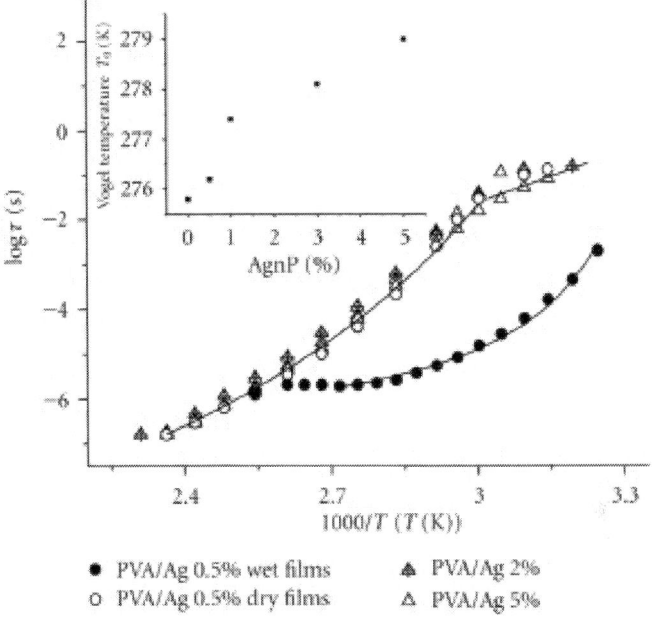

Figure 6: Relaxation time (τ) versus 1000/T (T in K).

The α-relaxation process is strongly dependent on moisture content since for the conductivity dependence on temperature, the relaxation time dependence for dry films is very different than that for wet films. After annealing, two well-defined behaviors are observed. A nonlinear behavior from 45°C to melting related to theα-relaxation process, and below 45°C an Arrhenius behavior associated to a secondary relaxation process which will be discussed later.

The results suggest a plasticizing effect of water on PVA composites in agreement with the results previously reported for pristine PVA [19]. A plasticized T_g can be observed in wet films through a nonlinear behavior described by the VFT model separated by the water evaporation region from a second nonlinear behavior disclosed at higher temperature. However, once water is eliminated by thermal treatment, this plasticized T_g vanished and only one nonlinear behavior is observed from 45°C to melting. Also the relaxation time for wet films is lower since the segmental motion is favored by water, by increasing the free volume between chains and because the orientation polarization of polar molecules is slowed down. However, this plasticizing effect is less evident in PVA composites when compared to pristine PVA with their water content being lower.

More information about moisture effect is provided by means of the complex modulus M^*, specifically by the M'' versus Temperature plot for wet and dry PVA and PVA composites as it is shown in Figure 7. In wet samples, two relaxations peaks (one below 50°C and another one above 80°C) at 5 kHz were found. As frequency increases, both peaks shift towards higher temperatures. These two relaxation peaks have been previously reported by other authors from M'', ε'', and tan δ plots for pure PVA, PVA blends, and composites [23, 25, 32–34]. These relaxations were previously assigned as the β-relaxation ascribed to side-group dipoles orientation (relaxation below 80°C, T_β= 40°C for pristine PVA [not shown] and 46°C for PVA/AgnP composite) and the glass transition of PVA (ca. T_g=85°C), respectively [23, 25, 27, 35, 36]. Nonetheless, our results show that for dry samples the lower temperature relaxation peak (below 50°C) vanishes after annealing the films at 120°C, remaining the α-relaxation peak at the same temperature for each frequency. This fact leads to the conclusion that the low temperature relaxation in wet samples can be traced to a moisture effect, and it does not correspond to a local mode β-relaxation. In this study, for pristine PVA dry samples T_g is ca. 88°C in agreement with previous results [19]. From Figure 7, it can also be observed that T_g increases with AgnP content; for PVA with AgnP 3 wt%, T_g is ca. 92°C. T_g values of composites and pristine PVA, calculated from the dielectric modulus plots as a function of AgnP content in dry films, are shown in Figure 7. It could be observed, that these T_g values are very different from those calculated by the empirical rule. Based upon our previous studies on pristine PVA [19] and chitin and chitosan Nano composites [20–23], we propose to use the Tg analysis based upon the modulus rather than the use of the empirical rule.

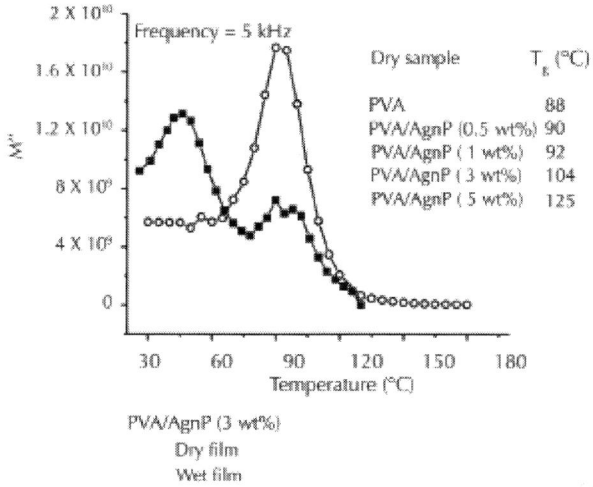

Figure 7: Imaginary part (*M"*) of the dielectric modulus (*M**) versus Temperature (°C) for PVA/AgnP (3%wt) wet and dry films.

PVA and moisture interaction would correspond to the formation of hydrogen bonds between PVA hydroxyl groups and water, and hydrogen bonds are the dominant interaction responsible for the structure as well as its molecular dynamics. It is possible that the interaction between OH groups and moisture is capable to destroy inter- and intrachain bonding in PVA affecting its crystalline regions; therefore, water acts as a plasticizer by an increase of the free volume in the amorphous phase [37]. This OH group's moisture interaction is disturbed by AgnP inclusion, and as a consequence, the molecular dynamics of both relaxation processes is strongly affected. The low temperature relaxation shifts to higher temperature as AgnP content increases resulting from lower moisture content and lower mobility.

On the other hand, regarding the behavior disclosed in dry films below 45°C previously observed in Figures 5and 6 (conductivity and relaxation time dependences), it could be seen that this relaxation follows a single Arrhenius behavior, which corresponds to a secondary relaxation process. In glassy polymers, the chains are frozen and molecular motions involved in secondary relaxations occur at higher time scale (i.e., higher relaxation times) as it is observed in Figure 6. This linear dependence can be explained based on the variable range hopping model. For most polymers this dependence of the dc

Conductivity on temperature (T) is often represented by the variable range hopping (VRH) model proposed by Mott [38, 39]:

$$\sigma_{dc}(T) = \sigma_0 \exp\left[-\left(\frac{T_0}{T}\right)^\gamma\right],$$

(2)

where σ_0 can be considered as the limiting value of conductivity at infinite temperature and $\sigma_0 \sim 1/T^{1/2}$ [40], T_0 depends on the localization and density of the states, and the exponent γ is related to the dimensionality d of the transport process via the equation $\gamma=1/(1+d)$, where d = 1, 2, 3. The applicability of the VRH model is examined by plotting the experimental results in the form of $\log(T)^{1/2}$ versus $T^{-\gamma}$ [39].

The experimental data in the 0°C to 45°C temperature range plotted according to the VRH model are presented on Figure 8. It is noteworthy that the dependence $\log (T)^{1/2}$ versus $T^{-\gamma}$ can be linearly fitted with both $\gamma=1/4$ and 1/3. However, the best least-square fitting is obtained for $\gamma=1/4$ (with $R^2= 0.994$). This value corresponds to a three-dimensional transport process as explained before.

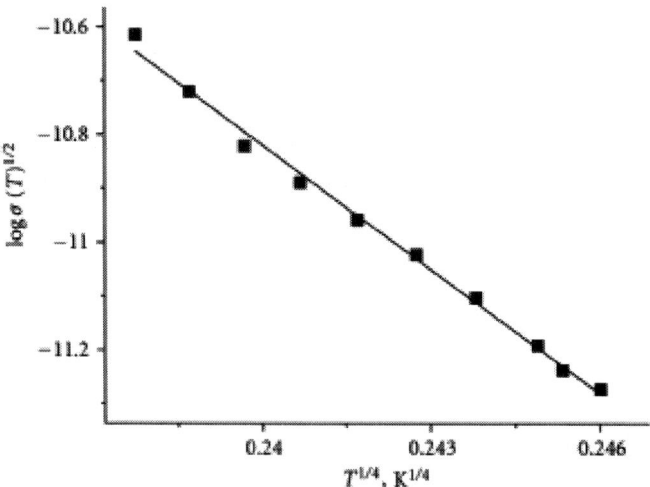

Figure 8: $\log \sigma T^{1/2}$ versus $T^{1/4}$ for pure PVA.

Furthermore, Linares et al. [28] reported the same linear behavior below 21°C for PVDF/PVA blends with different weight percent ratios. They ascribed this low temperature relaxation to a subglass relaxation process occurring in the amorphous phase and associated it to the polar groups attached to the polymeric chain. However, in our case, the application of the variable range hopping model is more appropriate to describe the AgnP-PVA system, since the inclusion of metal nanoparticles give rise to the formation of charge transfer complexes (CTCs) causing the reduction of the crystalline-amorphous interface, decreasing the interfacial barrier and increasing the transition probability of electron hopping across the barrier and insulator chains, which in turn provides a conducting path through the amorphous regions of the polymer matrix resulting in enhanced conductivity [16].

Dynamic Mechanical Analysis (DMA)

Gautam and Ram [33] report the preparation and thermomechanical properties of Ag-PVA nanocomposites; however, they do not discuss the strong influence of water content on the composite's relaxation processes. In the mentioned study, it is evident that authors assigned the glass transition temperature of PVA/AgnP composites dismissing the effect of water on the relaxation process of composites since the reported values are below 40°C which correspond to the plasticized T_g.

DMA analysis performed at 1 Hz in pristine PVA, and its composites wet and dry films are shown in Figure 9. All samples showed similar behaviors before and after being annealed. It can be observed that the tan δ peak in wet films shows a relaxation process near 40°C. This low temperature DMA peak was previously assigned several times as the glass transition temperature of pristine PVA and several PVA composites. Gautam and Ram [33] assigned T_g values between 40°C and 36°C in Ag-PVA nanocomposites depending upon Ag content. Tian and Tagaya [41] observed this peak around 50°C and described it as the glass transition of pristine PVA. They also show peaks for perlite/PVA and OMMT/PVA nanocomposites without taking into account water content. Yang et al. [42] assigned 44°C and 45°C values for the glass transition temperatures of pristine PVA and PVA/10 wt% montmorillonite (MMT) composites, respectively. They discussed that the T_g values from the DMA analyses for polymer membranes were lower than those from the DSC analyses (71–82°C) because the sensitivity for the measurement of a glass transition temperature by DMA is more sensitive than that by DSC.

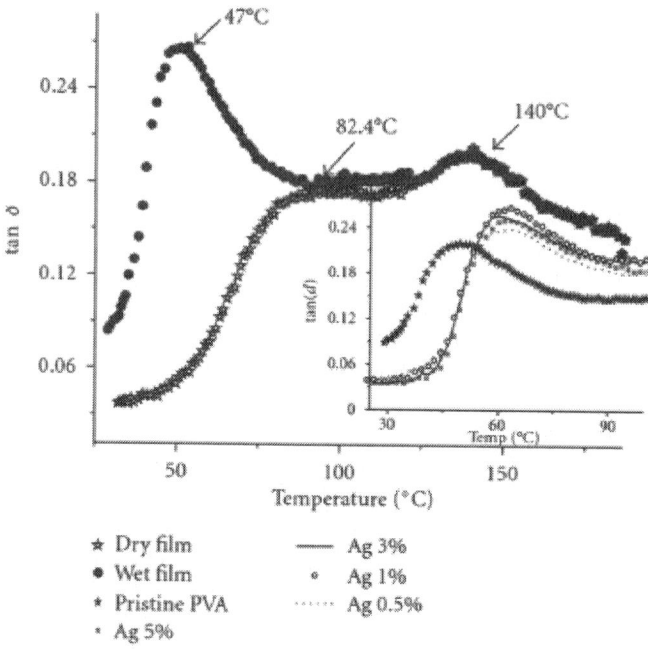

* Dry film   —— Ag 3%
* Wet film   ∘ Ag 1%
* Pristine PVA   ······ Ag 0.5%
* Ag 5%

Figure 9: tan δ versus temperature for dry and wet pristine PVA film. Window insert: tanδ versus temperature for wet pristine PVA and wet PVA/AgnP composites.

However, it is noteworthy that this tanδpeak vanishes in dry films after annealing at 120°C; consequently, this peak is related to water-polymer motions which mask the glass transition in wet films. After water evaporation, dry films disclose the glass transition event above 80°C.

Window insert of Figure 9 shows the variation of tanδpeak in wet PVA and PVA/AgnP composites films (water content around 4.1% calculated by TGA). Once again the influence of AgnP on PVA moisture absorption capacity is observed, that is, water-polymer motion is restricted by the presence of silver nanoparticles, which reduces moisture content in composites shifting the relaxation process from 47°C for pristine PVA to 63.8°C for the higher concentration of AgnP (5 wt%).

The storage modulus is shown in Figure 10, it can be seen that it increases as AgnP content in films increases up to 3 wt% suggesting significant reinforcement effect of Ag nanoparticles. Nanoparticles restrict the polymer's chain mobility due to their large surface area and their van der Waals attraction to the polymer matrix, causing the strengthening of mechanical properties of nanocomposite films [33]. However, at 5 wt% of AgnP, the storage modulus decreases compared to lower concentrations, but it remains higher than pristine PVA.

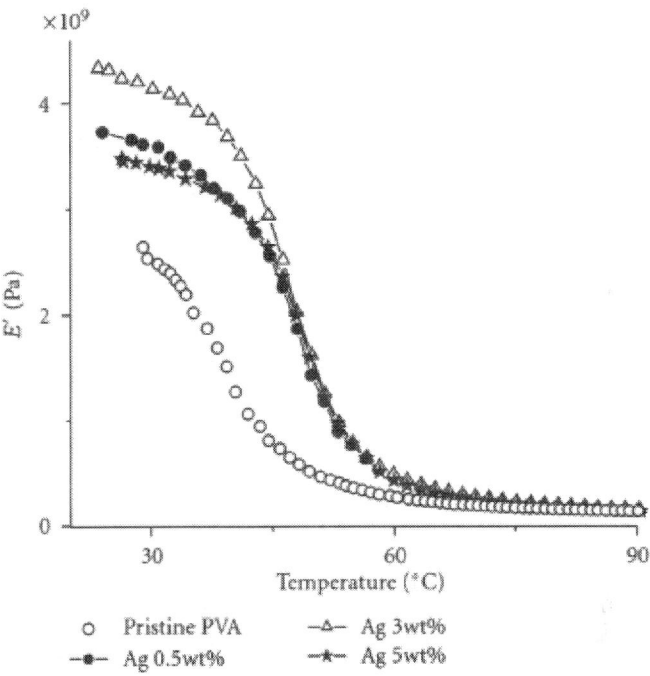

Figure 10: DMA storage modulus (E') versus temperature for wet pristine PVA and wet PVA/AgnP composites.

Gautam and Ram [33] showed this reinforcement up to 0.2 wt% of AgnP and then decreases for higher concentrations, being even lower than that of pristine PVA at concentration above 1 wt%. This decrease in the storage modulus is ascribed to the formation of large agglomerates, which in turns affects moisture content of PVA. The interaction between OH groups of PVA and moisture affects the crystalline regions. This OH groups moisture interaction is disturbed by AgnP inclusion resulting in lower moisture content and lower mobility; thus the strengthening of AgnP is less effective at higher concentrations of nanoparticles.

CONCLUDING REMARKS

The molecular dynamics of PVA/AgnP composites was studied by impedance spectroscopy in the 0.1 Hz to 110 MHz and 20°C to 300°C frequency and temperature ranges, respectively. As well as for pristine PVA, improper water elimination analysis in PVA composites could lead to misinterpretation of thermal relaxations in PVA composites such that a plasticized T_g for wet films has been assigned as a secondary β-relaxation in a number of previous studies in the literature.

Two well-defined nonlinear regions at low and high temperatures, with an intermediate discontinuity between 80°C and 100°C associated to moisture evaporation, were observed in wet PVA/AgnP composites films in agreement with pristine PVA behavior. Previously, these two regions were erroneously assigned to two Arrhenius-type relaxations in pristine and doped PVA. The evaporation of water and its plasticizing effect are more evident in pure PVA confirming the existence of strong interaction between OH groups of PVA chains and AgnP, as shown by FTIR analysis. Dry films show a single nonlinear VFT dependence (from 45°C until melting) associated to the α-relaxation and, therefore, to the glass transition phenomenon. T_g of composites increases as AgnP content increases from 88°C for pristine PVA to 125°C for PVA/AgnP (5 wt%). Below 45°C, dry films exhibit a single Arrhenius behavior showing a 3D hopping conductivity as explained based on the variable range hopping model.

PVA/AgnP composites have higher conductivity compared to pristine PVA, and it increases as AgnP weight percent increases. The inclusion of metal nanoparticles decreases the interfacial barrier and increases the transition probability of electron hopping across the barrier and insulator chains, which in turn provides a conducting path through the amorphous regions of the polymer matrix resulting in enhanced conductivity.

Finally, DMA measurements support the statement that a secondary relaxation was erroneously assigned as the glass transition of PVA and composites in previous reports.

ACKNOWLEDGMENTS

The authors thank José Alfredo Muñoz, Araceli Mauricio, and María del Carmen Díaz for their help provided. They also thank CONACYT (Mexican Government) for partial financial support. The authors are not partially or fully associated with Nanotechnologies nor do they endorse the use of their products.

REFERENCES

1. A. Henglein, "Small-particle research: physicochemical properties of extremely small colloidal metal and semiconductor particles," Chemical Reviews, vol. 89, no. 8, pp. 1861–1873, 1989.

2. R. Chapman and P. Mulvaney, "Electro-optical shifts in silver nanoparticle films," Chemical Physics Letters, vol. 349, no. 5-6, pp. 358–362, 2001. ·

3. L. N. Lewis, "Chemical catalysis by colloids and clusters," Chemical Reviews, vol. 93, no. 8, pp. 2693–2730, 1993.

4. A. Kiesow, J. E. Morris, C. Radehaus, and A. Heilmann, "Switching behavior of plasma polymer films containing silver nanoparticles," Journal of Applied Physics, vol. 94, no. 10, pp. 6988–6990, 2003. · ·

5. Y. Min, M. Akbulut, K. Kristiansen, Y. Golan, and J. Israelachvili, "The role of interparticle and external forces in nanoparticle assembly," Nature Materials, vol. 7, no. 7, pp. 527–538, 2008. · ·

6. J. L. Elechiguerra, J. L. Burt, J. R. Morones et al., "Interaction of silver nanoparticles with HIV-1,"Journal of Nanobiotechnology, vol. 3, article 6, 2005. · ·

7. A. L. Stepanov, V. N. Popok, I. B. Khaibullin, and U. Kreibig, "Optical properties of polymethylmethacrilate with implanted silver nanoparticles," Nuclear Instruments and Methods in Physics Research, Section B, vol. 191, no. 1–4, pp. 473–477, 2002. · ·

8. Y. Badr and M. A. Mahmoud, "Enhancement of the optical properties of poly vinyl alcohol by doping with Silver

nanoparticles," Journal of Applied Polymer Science, vol. 99, no. 6, pp. 3608–3614, 2006. · ·

9.   P. C. Lebaron, Z. Wang, and T. J. Pinnavaia, "Polymer-layered silicate nanocomposites: an overview,"Applied Clay Science, vol. 15, no. 1-2, pp. 11–29, 1999. · ·

10.   J. M. Yeh, S. J. Liou, C. Y. Lin, C. Y. Cheng, Y. W. Chang, and K. R. Lee, "Anticorrosively enhanced PMMA-clay nanocomposite materials with quaternary alkylphosphonium salt as an intercalating agent," Chemistry of Materials, vol. 14, no. 1, pp. 154–161, 2002. · ·

11.   T. H. Kim, L. W. Jang, D. C. Lee, H. J. Choi, and M. S. John, "Synthesis and rheology of intercalated polystyrene/Na⁺-Montmorillonite nanocomposites," Macromolecular Rapid Communications, vol. 23, no. 3, pp. 191–195, 2002. ·

12.   Y. J. Lee and W. S. Lyoo, "Preparation of Atactic poly(vinyl alcohol)/silver composite nanofibers by electrospinning and their characterization," Journal of Applied Polymer Science, vol. 115, no. 5, pp. 2883–2891, 2010. · ·

13.   V. K. Sharma, R. A. Yngard, and Y. Lin, "Silver nanoparticles: green synthesis and their antimicrobial activities," Advances in Colloid and Interface Science, vol. 145, no. 1-2, pp. 83–96, 2009. · ·

14.   K. H. Hong, J. L. Park, I. N. Hwan Sul, J. H. Youk, and T. J. Kang, "Preparation of antimicrobial poly(vinyl alcohol) nanofibers containing silver nanoparticles," Journal of Polymer Science, Part B, vol. 44, no. 17, pp. 2468–2474, 2006. · ·

15.   Z. H. Mbhele, M. G. Salemane, C. G. C. E. van Sittert, J. M. Nedeljković, V. Djoković, and A. S. Luyt, "Fabrication and characterization of silver-polyvinyl alcohol nanocomposites," Chemistry of Materials, vol. 15, no. 26, pp. 5019–5024, 2003. · ·

16.   S. Mahendia, A. K. Tomar, and S. Kumar, "Electrical conductivity and dielectric spectroscopic studies of PVA-Ag nanocomposite films," Journal of Alloys and Compounds, vol. 508, no. 2, pp. 406–411, 2010. · ·

17.   W. Brostow, R. Chiu, I. M. Kalogeras, and A. Vassilikou-Dova, "Prediction of glass transition temperatures: binary blends and copolymers," Materials Letters, vol. 62, no. 17-18, pp. 3152–3155, 2008. · ·

18. N. R. Jadhav, V. L. Gaikwad, K. J. Nair, and H. M. Kadam, "Glass transition temperature: basics and application in pharmaceutical sector," Asian Journal of Pharmaceutics, vol. 3, no. 2, pp. 82–89, 2009. · ·

19. J. B. González-Campos, Z. Y. García-Carvajal, E. Prokhorov, et al., "Revisiting thermal relaxations of poly(vinyl alcohol)," Journal of Applied Polymer Science, vol. 125, pp. 4082–4090, 2012. ·

20. J. B. Gonzalez-Campos, E. Prokhorov, G. Luna-Bárcenas et al., "Relaxations in chitin: evidence for a glass transition," Journal of Polymer Science, Part B, vol. 47, no. 9, pp. 932–943, 2009. · ·

21. J. B. Gonzalez-Campos, E. Prokhorov, G. Luna-Bárcenas, A. Fonseca-García, and I. C. Sanchez, "Dielectric relaxations of chitosan: the effect of water on the α-relaxation and the glass transition temperature," Journal of Polymer Science, Part B, vol. 47, no. 22, pp. 2259–2271, 2009. View at Publisher· ·

22. J. B. Gonzalez-Campos, E. Prokhorov, G. Luna-Bárcenas et al., "Chitosan/silver nanoparticles composite: molecular relaxations investigation by dynamic mechanical analysis and impedance spectroscopy," Journal of Polymer Science, Part B, vol. 48, no. 7, pp. 739–748, 2010. · ·

23. P. B. Bhargav, B. A. Sarada, A. K. Sharma, and V. V. R. N. Rao, "Electrical conduction and dielectric relaxation phenomena of PVA based polymer electrolyte films," Journal of Macromolecular Science, Part A, vol. 47, no. 2, pp. 131–137, 2010. · ·

24. A. L. Lee and A. J. Wand, "Microscopic origins of entropy, heat capacity and the glass transition in proteins," Nature, vol. 411, no. 6836, pp. 501–504, 2001. · ·

25. T. A. Hanafy, "Dielectric relaxation and alternating-current conductivity of gadolinium-doped poly(vinyl alcohol)," Journal of Applied Polymer Science, vol. 108, no. 4, pp. 2540–2549, 2008. · ·

26. S. L. Agrawal and A. Awadhia, "DSC and conductivity studies on PVA based proton conducting gel electrolytes," Bulletin of Material Science, vol. 27, no. 6, pp. 523–527, 2004. ·

27. A. Linares, A. Nogales, D. R. Rueda, and T. A. Ezquerra, "Molecular dynamics in PVDF/PVA blends as revealed by dielectric loss spectroscopy," Journal of Polymer Science, Part B, vol. 45, no. 13, pp. 1653–1661, 2007. · ·

28. M. Hema, S. Selvasekerapandian, and G. Hirankumar, "Vibrational and impedance spectroscopic analysis of poly(vinyl alcohol)-based solid polymer electrolytes," Ionics, vol. 13, no. 6, pp. 483–487, 2007. · ·

29. J. Zhang, M. Mine, D. Zhu, and M. Matsuo, "Electrical and dielectric behaviors and their origins in the three-dimensional polyvinyl alcohol/MWCNT composites with low percolation threshold," Carbon, vol. 47, no. 5, pp. 1311–1320, 2009. · ·

30. F. Garcia, A. Garcia-Bernabe, V. Compan, R. Diaz-Calleja, J. Guzman, and E. Riande, "Relaxation behavior of acrylate and methacrylate polymers containing dioxacyclopentane rings in the side chains,"Journal of Polymer Science B, vol. 39, no. 3, pp. 286–299, 2001. ·

31. V. Compan, J. Guzman, R. Diaz-Calleja, and E. Riande, Relxation behavior of methacrylic polymers with bulky hydrophilic groups in their structuresJournal of Polymer Science B, vol. 37, pp. 3027–3037, 1999.

32. G. G. Raju, Dielectrics in Electrical Fields, Marcel Dekker, New York, NY, USA, 2003.

33. A. Gautam and S. Ram, "Preparation and thermomechanical properties of Ag-PVA nanocomposite films," Materials Chemistry and Physics, vol. 119, no. 1-2, pp. 266–271, 2010. · ·

34. M. Köhler, P. Lunkenheimer, and A. Loidl, "Dielectric and conductivity relaxation in mixtures of glycerol with LiCl," European Physical Journal E, vol. 27, no. 2, pp. 115–122, 2008. · ·

35. M. D. Migahed, N. A. Bakr, M. I. Abdel-Hamid, O. El-Hanafy, and M. El-Nimr, "Dielectric relaxation and electric modulus behavior in poly(vinyl alcohol)-based composite systems," Journal of Applied Polymer Science, vol. 59, no. 4, pp. 655–662, 1996.

36. K. P. Singh and P. N. Gupta, "Study of dielectric relaxation in polymer electrolytes," European Polymer Journal, vol. 34, no. 7, pp. 1023–1029, 1998.

37. M. C. Hernández, N. Suárez, L. A. Martínez, J. L. Feijoo, S. Lo Mónaco, and N. Salazar, "Effects of nanoscale dispersion in the dielectric properties of poly(vinyl alcohol)-bentonite nanocomposites,"Physical Review E, vol. 77, no. 5, Article ID 051801, 2008. · ·

38. N. F. Mott, Metal-Insulator Transitions, Taylor & Francis, London, UK, 1990.

39. G. C. Psarras, "Hopping conductivity in polymer matrix-metal particles composites," Composites Part A, vol. 37, no. 10, pp. 1545–1553, 2006. · ·

40. S. Capaccioli, M. Lucchesi, P. A. Rolla, and G. Ruggeri, "Dielectric response analysis of a conducting polymer dominated by the hopping charge transport," Journal of Physics Condensed Matter, vol. 10, no. 25, pp. 5595–5617, 1998. · ·

41. H. Tian and H. Tagaya, "Dynamic mechanical property and photochemical stability of perlite/PVA and OMMT/PVA nanocomposites," Journal of Materials Science, vol. 43, no. 2, pp. 766–770, 2008. · ·

42. C. C. Yang, Y. J. Lee, and J. M. Yang, "Direct methanol fuel cell (DMFC) based on PVA/MMT composite polymer membranes," Journal of Power Sources, vol. 188, no. 1, pp. 30–37, 2009. · ·

Ecofriendly Synthesis of Anisotropic Gold Nanoparticles: A Potential Candidate of SERS Studies

Ujjwala Gaware, Vaishali Kamble,
and Balaprasad Ankamwar

Bioinspired Materials Science Laboratory, Department of Chemistry, University of Pune, Ganeshkhind, Pune 411007, India

ABSTRACT

Ecofriendly synthesis of nanoparticles has been inspiring to nanotechnologists especially for biomedical applications. Moreover, anisotropic particle synthesis is an attractive option due to decreased symmetry of such particles often leads to new and unusual chemical and physical behaviour. This paper reports a single-step room-temperature synthesis of gold nanotriangles using a cheap bioresource of reducing and stabilizing agentPiper betle leaf extract. On treating aqueous chloroauric acid solution with Piper betle leaf extract, after 12 hr, complete reduction of the chloroaurate ions was observed leading to the formation of flat and single crystalline gold nanotriangles. These gold nanotriangles can be exploited in photonics, optical coating, optoelectronics, magnetism, catalysis, chemical sensing, and so forth, and are a potential candidate of SERS studies.

INTRODUCTION

The synthesis of metal and semiconductor nanoparticles has innumerable opportunities of research due to their present and future applications in biosensing [1], chemical sensing [2], recording media [3], optoelectronics [4], and catalysis [5]. Masatake has reported [6] gold as a novel catalyst in the 21st century: it's preparation, working mechanism, and application as the catalyst in CO oxidation. The majority of earlier research has focused on isotropic, that is, spherical particles. However, anisotropic particles are particularly interesting because the decreased symmetry of such particles often leads to new and unusual chemical and physical properties [7]. In this context, ecofriendly biosynthesis protocols are better roadmap to avoid adverse effects of nanomaterials especially in medical applications. Moreover, use of plant extracts as a reducing and capping agent for the synthesis of nanoparticles could be advantageous over other environmentally benign biological processes by eliminating the elaborate process of maintaining cell cultures. It can also be suitably scaled up for the large-scale synthesis of nanoparticles. The biosynthesis of platinum nanoparticles usingDiospyros kaki leaf extract [8], silver nanoparticles using leaf extracts [9], silver and gold nanoparticles using phyllanthin [10], Clove extract [11], and within live Alfalfa plants in solid media [12] has been demonstrated. In recent studies, we have reported on the synthesis of gold nanoparticles by the reduction of aqueous $AuCl_4^-$ ions using Cymbopogon flexuosus [13], Tamarindus indica [2], Emblica officinalis [14], Terminalia catappa[15], Murraya koenigii and Citrus limonum leaf extracts [16].

In the case of bioreduction of aqueous gold ions by lemongrass extract, we did observe the formation of a large percentage of single crystalline, highly (111)-oriented gold nanotriangles with interesting optical absorption in the near infrared region of the electromagnetic spectrum [13]. Preliminary studies on the lemongrass extract and the gold nanotriangles indicated that ketones/aldehydes present in the extract may play an important role in directing the shape evolution in these nanostructures [13]. In order to test whether this hypothesis is true, we have looked at the composition of other plants for possible presence of such molecules and have identified the Piper betle plant as a potential candidate for shape-controlled synthesis of gold nanoparticles. This paper elaborates below the reaction of aqueous chloroaurate ions with Piper betle leaf extract results in the formation of single crystalline sharp vertices and truncated flat gold nanotriangles in large percentage with little number of hexagons. The edge-length of the nanotriangles varies from 660 to 1000 nm as would be expected from the highly anisotropic nature of the nanotriangles; they exhibit large absorption in the near infrared region. Surface-enhanced Raman spectroscopy (SERS) technique is generally used to enhance Raman signals by factors of 10^4–10^6 [17–19] for detecting the molecules at low concentrations and to acquire information of the surface of materials. Sabur et al. [20] demonstrated glycine detection limit as low as 10^{-12} M using SERS intensity optimization by controlling the size and shape of faceted gold nanoparticles. Moreover, Krpeti et al. [21] demonstrated that gold nanoparticles of different core sizes play an important role in the design of functional nanoparticles for colorimetric and SERS-based sensing applications, allowing controlled nanoparticles assembly and tunable sensor response for trace detection of Ni (II) ions in an aqueous solution. The recent findings used to improve the gold nanoparticle-based SERS substrates with ultrahigh sensitivity for the detection of bacterial spores [22] and gold nanoparticle-coated biomaterial as SERS microprobes [23]. These reports suggest that the ecofriendly synthesized anisotropic gold nanoparticles could be a potential candidate of SERS studies. Presented below are details of the investigation.

MATERIALS AND METHODS

Biosynthesis of Anisotropic Gold Nanoparticles

The broth used for the reduction of Au^{3+} ions to Au^0 was prepared by taking 24 g of thoroughly washed and finely cut Piper betle leaves in a 500 mL Erlenmeyer flask with 100 mL of sterile distilled water. This mixture was then boiled for 5 min and filtered through four-fold muslin cloth on cooling to room temperature. In a typical experiment, 4 mL of this broth was added to 90 mL of 1×10^{-3} M aqueous chloroauric acid ($HAuCl_4$) solution at room temperature. The bioreduction of chloroaurate ions in solution was monitored by periodic sampling and measuring the UV-Vis-NIR spectra of the solutions.

UV-Vis-NIR Spectroscopy Studies

UV-vis-NIR spectroscopy measurements of the Piper betle leaves extract-reduced gold nanotriangles were carried out as a function of time of reaction at room temperature on a JASCO dual-beam spectrophotometer (model V-570) operated at a resolution of 1 nm.

X-Ray Diffraction (XRD) Measurement

X-ray diffraction measurement of gold nanotriangles powder was carried out on a Bruker axs (model D8 Advance) instrument operating at a voltage of 40 kV and current of 40 mA with Cu K radiation.

Fourier Transform Infrared (FTIR) Spectroscopy Measurements

After complete reduction of $AuCl_4^-$ ions by the Piper betle leaves extract, for the isolation of gold nanoparticles from the free proteins or other organic biomolecular compounds existing in the solution, the centrifugation was carried out at 6000 rpm for 15 min. Thus, obtained gold nanoparticle pellets after centrifugation were redispersed in water prior to FTIR analysis. Films of the purified gold nanoparticles were deposited on Si (111) wafers by simple drop coating and were subjected to FTIR analysis on a Perkin-Elmer FTIR Spectrum One spectrophotometer in the diffuse reflectance mode at a resolution of 4 cm^{-1}.

Transmission Electron Microscopy (TEM) Measurements

TEM samples of the gold nanotriangles were prepared by placing a drop of the nanoparticle solution on carbon-coated copper grids and allowing the solvent to evaporate; TEM measurements were performed on a JEOL model 1200EX instrument operated at an accelerating voltage at 120 kV.

RESULTS AND DISCUSSION

The kinetics of reduction of aqueous chloroaurate ions during reaction with the Piper betle leaves broth was followed by UV-vis-NIR spectroscopy. It is well known that gold nanoparticles exhibit various shades of colors depending on size and shapes of nanoparticles, which also supports the coffee color observed in this study, which arises due to excitation of surface plasmon resonance (SPR) in the gold nanoparticles [24]. Figure 1(a) shows the UV-vis-NIR spectra recorded from the aqueous chloroauric acid Piper betle leaves broth reaction medium as a function of time of reaction. It is observed that as the reaction proceeds, the gold SPR band at ca. 551 nm steadily increases in intensity. This band is indicative of the presence of spherical nanoparticles in solution. In addition to the peak at 551 nm, a progressive increase in the absorption at longer wavelengths into the near-infrared (NIR) region of the electromagnetic spectrum is observed. The UV-vis-NIR spectra indicate that the peak in the long wavelength absorption is well into the NIR (Figure 1(a)). The spectrum of the biologically synthesized gold nanoparticles now clearly shows a peak centered at 1213 nm, after 12 hrs of the reaction. The long wavelength absorption both in solution could be either due to aggregation of spherical gold nanoparticles in solution [1, 13] or due to formation of anisotropic nanoparticles [25]. In our earlier work, we observed sintering of small spherical gold nanoparticles at room temperature to single crystalline gold nanotriangles, suggesting that the nanoparticle surface is liquid-like [13]. Figure 1(b) shows a TEM image recorded from the biologically synthesized gold nanoparticles at the end of the reaction with Piper betle leaves extract. The TEM image shows (Figure 1(b)) that the gold nanoparticles are predominantly triangular morphology. The biosynthesized nanotriangles consist of a mixture of triangles, truncated triangles, and hexagons. The truncation appears to be a common feature in such disk-like metal nanostructures and has been repeatedly observed in chemically prepared gold [26, 27] and silver nanotriangles [28, 29]. It is not clear what the reasons are for the formation of truncated nanotriangles. An analysis of the nanoparticles indicated that the percentage of gold nanotriangles/ hexagons in the as-prepared reaction medium was ca. 60% but could be enhanced to nearly 90% through two cycles of centrifugation at 6,000 rpm, washing and redispersion. The gold nanotriangles in high magnified TEM image of Figure 1(d) shows considerable contrast along their surface. This contrast is due to strains in the nanoparticles indicating that they are extremely flat, thin, and easily buckle [30]. The TEM analysis thus clearly clarifies that the strong absorption in the NIR observed for thePiper betle leaf extract prepared gold nanoparticles is due to the formation of highly anisotropic nanostructures and not due to assembly of spherical gold nanoparticles. Here the edge-length of the nanotriangles varies from 660 to 1000 nm (Figure 1(b)); nanoprism structures with edge lengths as large as several micrometers have been synthesized earlier, but these have not exhibited the optical or chemical properties associated with their smaller analogs [31–33]. Technically, triangular nanoprisms contain three sharp vertices that contribute significantly to their optical and electronic properties [31, 33]. In some cases, nanoprisms dimensions can be controlled in situ by adjusting experimental parameters, including metal ion and reducing agent ratios [34]. Earlier report has shown a mixture of spherical and polycrystalline nanoplatelet shapes with a low yield of either spherical or nanoplatelets [35]. In this paper bio-source of reducing and capping agent Piper betle leaves extract has resulted into high yield of formation of higher edge-length of single crystalline nanoprisms (660–1000 nm) compare to our earlier reports of gold nanoprisms synthesis using extracts of biosourcesCymbopogon flexuosus (200–500 nm) [13] and Tamarindus indica (100–500 nm) [2].

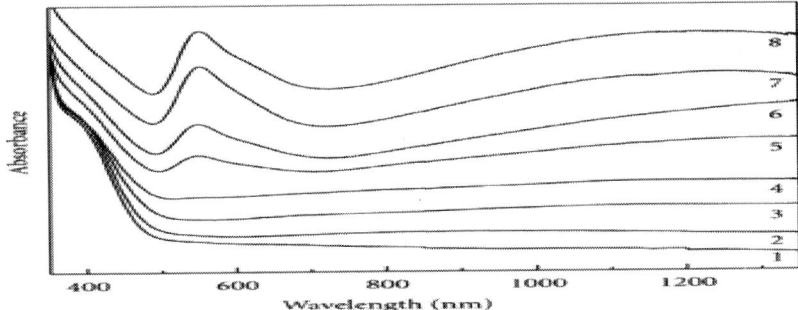

(a)

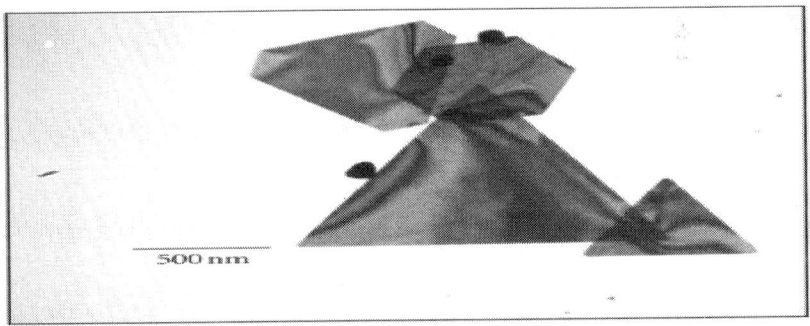

(b)

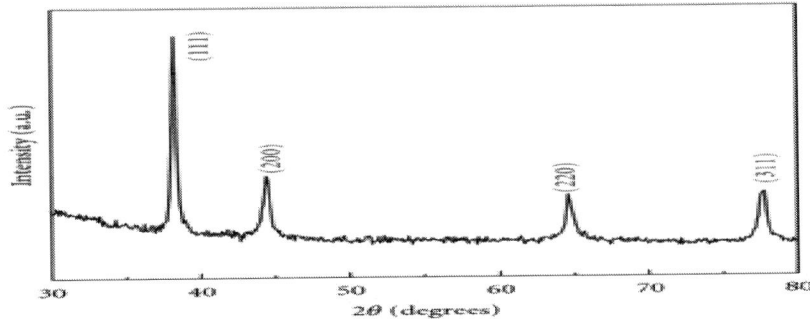

(c)

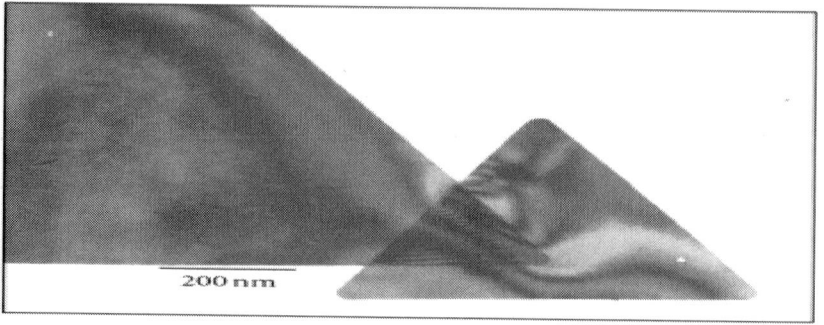

(d)

Figure 1: (a) UV-vis-NIR kinetics of the reaction of Piper betle leaf extract with aqueous chloroaurate ions. Curves 1–8 correspond to 0, 10, 60, 120, 180, 240, 480, and 720 min after reaction, respectively. (b) Representative TEM micrographs of triangular gold nanoparticles obtained by the reduction of $AuCl_4^-$ by Piper betle leaf extract. (c) XRD pattern of Piper betel leaf extract reduced nanotriangles with identified Bragg reflections. (d) A high magnified TEM micrograph of the single gold nanotriangles.

Figure 1(c) shows the X-ray diffraction patterns of the gold nanotriangles obtained in the Piper betle leaves extract reaction. The 2 values of the standard Au nanoparticles 38.184, 44.392, 64.576, and 77.547 degrees correspond to the Bragg reflections (111), (200), (220), and (311). Where as the 2 values of the obtained gold nanotriangles 38.2, 44.4, 64.6, and 77.8 degrees correspond to the Bragg reflections (111), (200), (220), and (311) that may be indexed on the basis of the fcc structure of gold. In consideration of wavelength of light source (= 1.54056 Å), our 2 values are almost matching with the JCPDF file no 04–0784 for gold. The (200), (220), and (311) Bragg reflections are extremely weak and considerably broadened relative to the intense (111) reflection. This interesting feature indicates that gold nanocrystals are highly anisotropic in nature and that the particles in the film are (111)-oriented.

FTIR measurements were carried out to identify the potential biomolecules in the Piper betle leaf broth responsible for the reduction of the chloroaurate ions and also the capping reagent responsible for the stability of the bioreduced gold nanoparticles. Bombay Piper betle leaves contain [36] reducing sugars (as glucose) 1.4–3.2%, non-reducing sugars (as sucrose) 0.6–2.5%, total sugars 2.4–5.6, Starch 1.0–1.2%, essential oil 0.8–1.8, and tannin 1.0–1.3. This composition data was used as a guideline to identify possible reducing and stabilizing biomolecules from the Piper betle leaves. Curves 1 and 2 of Figure 2(a) represent the FTIR spectrum of thePiper betle leaves extract and Piper betle leaves extract reduced gold nanoparticles with absorption bands at 759, 792, 812, 864, 914, 965, 996, 1115, 1147, 1282, 1410, 1514, 1602, and 3195 cm⁻¹ and 806, 912, 1018, 1110, 1263, 1384, 1522, and 2966 cm⁻¹, respectively.

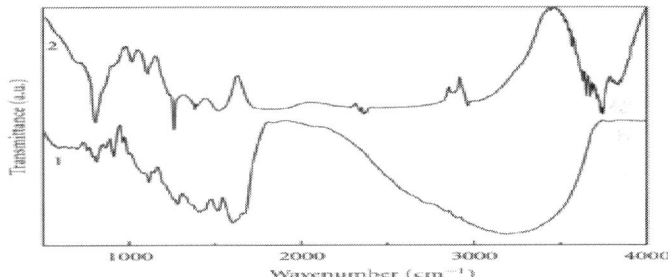

(a)

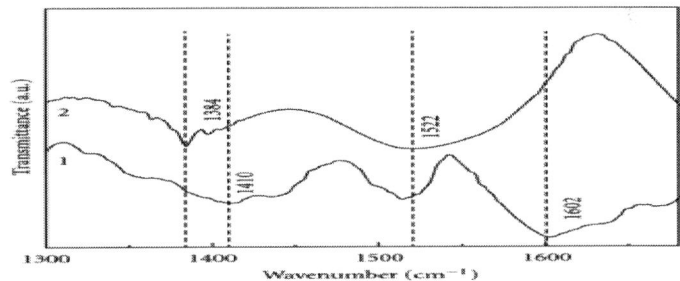

(b)

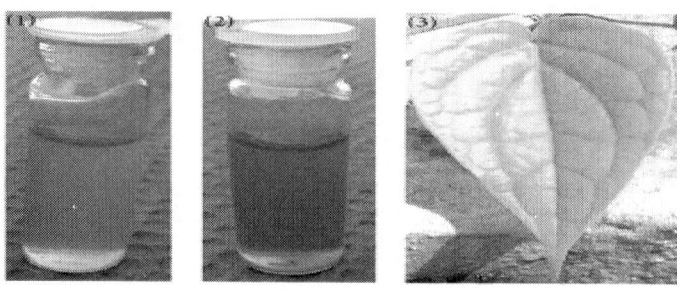

(c)

Figure 2: FTIR spectra of (1) pure Piper betle leaf broth and (2) Piper betle leaf-reduced gold nanoparticles in the wavenumbers range from (a) 400 to 4000 cm^{-1} and (b) 1300 to 1750 cm^{-1}; (c) it shows aqueous extract of Piper betle leaf (1), Piper betle leaf-reduced gold nanoparticles (2), and Piper betle leaf (3).

The shoulder at 1602 cm^{-1}, is characteristic of carbonyl stretch vibrations in ketones, aldehydes, and carboxylic acids. The 1616 cm^{-1} band is assigned to aromatic C–C skeletal vibrations/N–H deformations, most likely from indoleacetic acid [13]. Curve 2 shows the FTIR spectrum of the Piper betle leaves extract reduced gold with the absorption bands at 1602 cm^{-1} and 1410 cm^{-1}. The shift of the 1602 cm^{-1} band to 1522 cm^{-1} is attributed to binding of aldehydes/ketones with the gold nanoparticle surface [13]. The band at 1602 cm^{-1} is characteristics of carbonyl stretch vibrations [37], possibly from the acid groups present in the Piper betle leaf extract. The shift in the carbonyl stretch frequency (1602 cm^{-1}) to lower wave numbers (1522 cm^{-1}) followed by the disappearance of the 1602 cm^{-1} resonance may be due to its binding with the gold nanoparticle surface. A comparison of the two spectra also reveals the presence of prominent feature at ca. 3195 cm^{-1} in curve 1 while the 3195 cm^{-1} feature shifts to 2966 cm^{-1} due to coordination of the amine molecule with gold nanoparticles surface [38].

In our earlier report, we had mentioned that we believe the formation of gold nanotriangles is due to reduction of aqueous AuCl$_4^-$ ions by the reducing sugars, Raveendran et al. [39] used the reducing sugar -D-glucose as the reducing agent in the synthesis of green synthesis of silver nanoparticles. Bombay Piper betle leaves used in this paper contain reducing sugars (as glucose) 1.4–3.2% [36]. FTIR also supports the aldehydes/ketones bind to the nascent spherical nanoparticles rendering them "liquid-like" and amenable to sintering at room temperature as we earlier reported [13]. We did control experiments with different amounts of Piper betle leaf extract keeping constant volume and concentration of chorloauric acid; in this study, we observed that slow reduction rate was favoring the formation of anisotropic structures such as triangular structure. This could be achieved by maintaining appropriate concentration ratio of precursor and leaf extract. This study helps to elucidate the role of economical bio-resource, that is, aqueous plant extracts in shape-directing factors involved in ecofriendly syntheses of anisotropic metallic nanoparticles. Moreover, these anisotropic metal nanoparticles show "lightning-rod effect," another kind of field enhancement refers to enhanced charge density localization at a tip or vertex of nanoparticles. When an electromagnetic field (e.g., laser light) excites the free electrons of a metallic tip, a highly localized, strong electric field develops at these sharp tips or vertex with large curvatures, leading to a large field enhancement in those regions. This is the reason for the high Surface Enhanced Raman Scattering (SERS) activity of anisotropic nanoparticles. In our laboratory studies on the use of synthesized anisotropic gold nanoparticles as an effective SERS active substrate are going on which will be communicated later. This biogenic synthesis of metal nanoparticles is also important in bionanotechnology and can be utilized as a light energy conversion devices by combine with an organic dyad.

CONCLUSIONS

The one step synthesis of stable gold nanotriangles in high concentration using Piper betle leaves extract has been demonstrated. The reducing sugars (as glucose) are mainly responsible for formation of metallic gold nanoparticles. The high absorption coefficient of these gold triangles in the NIR region can be exploited in fabricating photonic devices such as optical sensors and in hyperthermia of tumors [40]. Gold nanotriangles have also the characteristics required for chemical sensor development [2]. In addition to chemical sensing the work of bio-sensing, SERS studies and catalysis are also the best candidatures for these anisotropic gold nanoparticles. Moreover, these structures are especially interesting because they have plasmonic features in the visible and IR regions, can be prepared in high yield and readily functionalized with a variety of sulfur-containing adsorbates [41–43], and is currently being pursued.

ACKNOWLEDGMENTS

The authors wish to express their appreciation to BCUD (BCUD/OSD/390; dated 16/11/2010), University of Pune for financial support.

REFERENCES

1. C. A. Mirkin, R. L. Letsinger, R. C. Mucic, and J. J. Storhoff, "A DNA-based method for rationally assembling nanoparticles into macroscopic materials," Nature, vol. 382, no. 6592, pp. 607–609, 1996.

2. B. Ankamwar, M. Chaudhary, and M. Sastry, "Gold nanotriangles biologically synthesized using tamarind leaf extract and potential application in vapor sensing," Synthesis and Reactivity in Inorganic, Metal-Organic, and Nano-Metal Chemistry, vol. 35, no. 1, pp. 19–26, 2005.

3. S. Sun, C. B. Murray, D. Weller, L. Folks, and A. Moser, "Monodisperse FePt nanoparticles and ferromagnetic FePt nanocrystal superlattices," Science, vol. 287, no. 5460, pp. 1989–1992, 2000.

4. D. H. Gracias, J. Tien, T. L. Breen, C. Hsu, and G. M. Whitesides, "Forming electrical networks in three dimensions by self-assembly," Science, vol. 289, no. 5482, pp. 1170–1172, 2000.

5. M. Valden, X. Lai, and D. W. Goodman, "Onset of catalytic activity of gold clusters on titania with the appearance of nonmetallic properties," Science, vol. 281, no. 5383, pp. 1647–1650, 1998.

6. H. Masatake, "Gold as a novel catalyst in the 21st century: Preparation, working mechanism and applications," Gold Bulletin, vol. 37, no. 1-2, pp. 27–36, 2004.

7. J. E. Millstone, S. J. Hurst, G. S. Métraux, J. I. Cutler, and C. A. Mirkin, "Colloidal gold and silver triangular nanoprisms," Small, vol. 5, no. 6, pp. 646–664, 2009.

8. J. Y. Song, E. Y. Kwon, and B. S. Kim, "Biological synthesis of platinum nanoparticles using Diopyros kaki leaf extractBioprocess and Biosystems Engineering," vol. 33, pp. 159–164, 2010.

9. J. Y. Song and B. S. Kim, "Rapid biological synthesis of silver nanoparticles using plant leaf extracts,"Bioprocess and Biosystems Engineering, vol. 32, pp. 79–84, 2009.

10. J. Kasthuri, K. Kathiravan, and N. Rajendiran, "Phyllanthin-assisted biosynthesis of silver and gold nanoparticles: a novel biological approach," Journal of Nanoparticle Research, vol. 11, no. 5, pp. 1075–1085, 2009.

11. A. K. Singh, M. Talat, D. P. Singh, and O. N. Srivastava, "Biosynthesis of gold and silver nanoparticles by natural precursor clove and their functionalization with amine group," Journal of Nanoparticle Research, vol. 12, no. 5, pp. 1667–1675, 2010

12. J. L. Gardea-Torresdey, E. Gomez, J. R. Peralta-Videa, J. G. Parsons, H. Troiani, and M. Jose-Yacaman, "Alfalfa sprouts: a natural source for the synthesis of silver nanoparticles," Langmuir, vol. 19, no. 4, pp. 1357–1361, 2003.

13. S. S. Shankar, A. Rai, B. Ankamwar, A. Singh, A. Ahmad, and M. Sastry, "Biological synthesis of triangular gold nanoprisms," Nature Materials, vol. 3, no. 7, pp. 482–488, 2004.

14. B. Ankamwar, C. Damle, A. Ahmad, and M. Sastry, "Biosynthesis of gold and silver nanoparticles using Emblica Officinalis fruit extract, their phase transfer and transmetallation in an organic solution,"Journal of Nanoscience and Nanotechnology, vol. 5, no. 10, pp. 1665–1671, 2005.

15. B. Ankamwar, "Biosynthesis of gold nanoparticles (Green-Gold) using leaf extract of Terminalia Catappa," E-Journal of Chemistry, vol. 7, no. 4, pp. 1334–1339, 2010.

16. B. Ankamwar, Biosynthesis: An Eco-Friendly Approach of Nanomaterials Synthesis, Chemical and Biomedical Applications, VDM, 2010.

17. P. Hildebrandt and M. Stockburger, "Surface-enhanced resonance Raman spectroscopy of Rhodamine 6G adsorbed on colloidal

silver," The Journal of Physical Chemistry B, vol. 88, no. 24, pp. 5935–5944, 1984.

18. X. M. Dou, Y. M. Jung, Z. Q. Cao, and Y. Ozaki, "Surface-enhanced raman scattering of biological molecules on metal colloid II: effects of aggregation of gold colloid and comparison of effects of ph of glycine solutions between gold and silver colloids," Applied Spectroscopy, vol. 53, no. 11, pp. 1440–1447, 1999.

19. H. X. Xu, J. Aizpurua, M. Kall, and P. Apell, "Electromagnetic contributions to single-molecule sensitivity in surface-enhanced Raman scattering," Physical Review E, vol. 62, pp. 4318–4324, 2000.

20. A. Sabur, M. Havel, and Y. Gogotsi, "SERS intensity optimization by controlling the size and shape of faceted gold nanoparticles," Journal of Raman Spectroscopy, vol. 39, no. 1, pp. 61–67, 2008.

21. Ž. Krpeti , L. Guerrini, I. A. Larmour, J. Reglinski, K. Faulds, and D. Graham, "Importance of nanoparticle size in colorimetric and sers-based multimodal trace detection of Ni(II) Ions with functional gold nanoparticles," Small, vol. 8, pp. 707–714, 2012.

22. H. W. Cheng and R. Q. Yu, "Nanoparticle-based substrates for surface-enhanced Raman scattering detection of bacterial spores," Analyst, vol. 137, pp. 3601–3608, 2012.

23. G. V. Pavankumar, "Gold nanoparticle-coated biomaterial as SERS micro-probes," Bulletin of Materials Science, vol. 34, no. 3, pp. 417–422, 2011.

24. P. Mulvaney, "Surface plasmon spectroscopy of nanosized metal particles," Langmuir, vol. 12, no. 3, pp. 788–800, 1996.

25. E. Hao, K. L. Kelly, J. T. Hupp, and G. C. Schatz, "Synthesis of silver nanodisks using polystyrene mesospheres as templates," Journal of the American Chemical Society, vol. 124, no. 51, pp. 15182–15183, 2002.

26. N. Malikova, I. Pastoriza-Santos, M. Schierhorn, N. A. Kotov, and L. M. Liz-Marzán, "Layer-by-layer assembled mixed spherical and planar gold nanoparticles: control of interparticle interactions,"Langmuir, vol. 18, no. 9, pp. 3694–3697, 2002.

27. Y. Shao, Y. Jin, and S. Dong, "Synthesis of gold nanoplates by aspartate reduction of gold chloride,"Chemical Communications, vol. 10, no. 9, pp. 1104–1105, 2004.

28. S. Chen and D. L. Carroll, "Synthesis and characterization of truncated triangular silver nanoplates,"Nano Letters, vol. 2, no. 9, pp. 1003–1007, 2002.

29. R. Jin, Y. Cao, C. A. Mirkin, K. L. Kelly, G. C. Schatz, and J. G. Zheng, "Photoinduced conversion of silver nanospheres to nanoprisms," Science, vol. 294, no. 5548, pp. 1901–1903, 2001.

30. W. T. S. Huck, N. Bowden, P. Onck, T. Pardoen, J. W. Hutchinson, and G. M. Whitesides, "Ordering of spontaneously formed buckles on planar surfaces," Langmuir, vol. 16, no. 7, pp. 3497–3501, 2000.

31. K. L. Kelly, E. Coronado, L. L. Zhao, and G. C. Schatz, "The optical properties of metal nanoparticles: the influence of size, shape, and dielectric environment," The Journal of Physical Chemistry B, vol. 107, no. 3, pp. 668–677, 2003.

32. Y. L. Luo, "Large-scale preparation of single-crystalline gold nanoplates," Materials Letters, vol. 61, no. 6, pp. 1346–1349, 2007.

33. K. L. Shuford, M. A. Ratner, and G. C. Schatz, "Multipolar excitation in triangular nanoprisms," The Journal of Chemical Physics, vol. 123, no. 11, pp. 114713–114722, 2005.

34. G. S. Métraux and C. A. Mirkin, "Rapid thermal synthesis of silver nanoprisms with chemically tailorable thickness," Advanced Materials, vol. 17, no. 4, pp. 412–415, 2005.

35. K. Sneha, M. Sathishkumar, S. Kim, and Y. S. Yun, "Counter ions and temperature incorporated tailoring of biogenic gold nanoparticles," Process Biochemistry, vol. 45, no. 9, pp. 1450–1458, 2010.

36. A. Krishnamurthi, Ed., The Wealth of India, A Dictionary of Indian Raw Materials & Industrial Products, vol. 8 of Raw Materials, National Institute of Science Communication, CSIR, New Delhi, India, 1998.

37. V. Patil, R. B. Malvankar, and M. Sastry, "Role of particle size in individual and competitive diffusion of carboxylic acid derivatized colloidal gold particles in thermally evaporated fatty amine films," Langmuir, vol. 15, no. 23, pp. 8197–8206, 1999.

38. D. V. Leff, L. Brandt, and J. R. Heath, "Synthesis and characterization of hydrophobic, organically-soluble gold

nanocrystals functionalized with primary amines," Langmuir, vol. 12, no. 20, pp. 4723–4730, 1996.

39. P. Raveendran, J. Fu, and S. L. Wallen, "Completely "green» synthesis and stabilization of metal nanoparticles," Journal of the American Chemical Society, vol. 125, no. 46, pp. 13940–13941, 2003.

40. L. R. Hirsch, R. J. Stafford, J. A. Bankson et al., "Nanoshell-mediated near-infrared thermal therapy of tumors under magnetic resonance guidance," Proceedings of the National Academy of Sciences of the United States of America, vol. 100, no. 23, pp. 13549–13554, 2003.

41. J. E. Millstone, S. Park, K. L. Shuford, L. Qin, G. C. Schatz, and C. A. Mirkin, "Observation of a quadrupole plasmon mode for a colloidal solution of gold nanoprisms," Journal of the American Chemical Society, vol. 127, no. 15, pp. 5312–5313, 2005.

42. C. Xue and C. A. Mirkin, "pH-switchable silver nanoprism growth pathways," Angewandte Chemie International Edition, vol. 46, no. 12, pp. 2036–2038, 2007.

43. R. Jin, Y. C. Cao, E. Hao, G. S. Métraux, G. C. Schatz, and C. A. Mirkin, "Controlling anisotropic nanoparticle growth through plasmon excitation," Nature, vol. 425, no. 6957, pp. 487–490, 2003.

Chapter 7

Detection Limits of DLS and UV-Vis Spectroscopy in Characterization of Polydisperse Nanoparticles Colloids

Emilia Tomaszewska[1], Katarzyna Soliwoda[1], Kinga Kadziola[1], Beata Tkacz-Szczesna[1], Grzegorz Celichowski[1], Michal Cichomski[1], Witold Szmaja[2], and Jaroslaw Grobelny[1]

[1]Department of Materials Technology and Chemistry, Faculty of Chemistry, University of Lodz, Pomorska 163, 90-236 Lodz, Poland

[2]Department of Solid State Physics, Faculty of Physics and Applied Informatics, University of Lodz, Pomorska 149/153, 90-236 Lodz, Poland

ABSTRACT

Dynamic light scattering is a method that depends on the interaction of light with particles. This method can be used for measurements of narrow particle size distributions especially in the range of 2–500 nm.

Sample polydispersity can distort the results, and we could not see the real populations of particles because big particles presented in the sample can screen smaller ones. Although the theory and mathematical basics of DLS technique are already well known, little has been done to determine its limits experimentally. The size and size distribution of artificially prepared polydisperse silver nanoparticles (NPs) colloids were studied using dynamic light scattering (DLS) and ultraviolet-visible (UV-Vis) spectroscopy. Polydisperse colloids were prepared based on the mixture of chemically synthesized monodisperse colloids well characterized by atomic force microscopy (AFM), transmission electron microscopy (TEM), DLS, and UV-Vis spectroscopy. Analysis of the DLS results obtained for polydisperse colloids reveals that several percent of the volume content of bigger NPs could screen completely the presence of smaller ones. The presented results could be extremely important from nanoparticles metrology point of view and should help to understand experimental data especially for the one who works with DLS and/or UV-Vis only.

INTRODUCTION

Nanoparticles (NPs) of noble metals, especially silver (Ag) and gold (Au) NPs, attract much attention in various fields. This is because of the unique properties of nanoscale objects, which are completely different from bulk materials, coatings, and single atoms. Applications or potential applications of NPs are not only diverse but also interdisciplinary and are related to material science, electronics, optics, and biomedicine. Noble metal NPs (in particular Ag NPs) can facilitate important advances in detection, diagnosis, and treatment of human cancers as well as delivery of drug and gene to malignant cells [1, 2]. By exploiting the unique optical and electronic properties of Au NPs, several new methods for ultrasensitive detection of DNA, RNA, and proteins have been developed [3–7].

Currently, many applications of silver NPs are related to their versatile antibacterial activity against a broad spectrum of bacteria without releasing toxic biocides. It should be noted that depending on the concentration and size silver nanoparticles can be considered as nontoxic as well as toxic [8–10]. Ag NPs are environmentally friendly antibacterial materials and therefore are used in many cosmetic products and medical applications such as hydrogel dressings [11]. Metal NPs are also promising for surface-enhanced Raman scattering (SERS) applications in detection and analysis of molecules. The enhancement factor can be strong enough that the technique allows detection of a single molecule [12–14].

As noble metal NPs are widely used in many applications, it is important to synthesize high-quality materials. In certain applications it is also important to prepare colloids with high concentrations. The size and size distribution are becoming extremely significant when quantum-sized effects are used to control material properties. Therefore, the control and complete analysis of the average particle size and narrow size distribution is essential to use NPs in many applications [15, 16].

There are a number of methods for nanoparticles size characterization, for example, scanning/transmission electron microscopy (SEM/TEM), atomic force microscopy (AFM) analytical ultracentrifugation (AUC), dynamic light scattering (DLS), and flow field fractionation (FFF). It is also possible to calculate particles size with the use of X-ray diffraction (XRD) patterns as well as with the shift of the band gap absorption in the UV-Vis spectrum [17, 18]. In case of the above-mentioned method [19–23], the particle size distribution is measured in dispersion or after drying the sample. The detailed description of the above-mentioned methods is beyond the scope of this work and will not be described.

There is a range of methods available for the particle size determination, and users should remember that precise characterization of particles requires the usage of few of them. Moreover, it is essential to know about the strengths and weaknesses of applied method for nanoparticles characterization in order to recognize the limits; for example, FFF and AUC methods suffer from artifacts like too small particle diameter because of the particle charge [24]. In case of TEM, it is not always obvious to recognize grain boundaries in aggregates. The main advantage of microscopic techniques is that it is possible to get the information about the morphology and the size of particles at the same time, but the preparation of samples for analysis is crucial (e.g., NPs must be deposited on the well-characterized substrate). Moreover, sample preparation is often time consuming, requires high precision and the use of appropriate reagents [25–27]. Microscopic techniques can also be problematic in case of the polydisperse samples because of possibility of particles aggregation or sample fractionation during drying.

Among the techniques of nanoparticles characterization the most commonly used are DLS [23, 28, 29] and UV-Vis spectroscopy [30–33]. The theory and mathematical basics of DLS and UV-Vis techniques are already well known [34, 35]. DLS measures the light scattered from the laser that passes through a colloid. Next, the modulation of the scattered light intensity as a function of time is analyzed, and the hydrodynamic size of particles can be determined [36, 37]. In case of UV-Vis spectroscopy, the intensity of light that is passing through the sample is measured. Nanoparticles have optical properties that are very sensitive on size, shape, agglomeration, and concentration changes. The unique optical properties of metal nanoparticles are a consequence of the collective oscillations of conduction electrons, which excited by electromagnetic radiation are called surface plasmon polariton resonances (SPPR) [35]. Those changes have an influence on the refractive index next to the nanoparticles surface; thus it is possible to characterize nanomaterials using UV-Vis spectroscopy. DLS and UV-Vis spectroscopy are fast and easy to operate techniques for particles characterization, especially for colloidal suspensions [38, 39]. There are several advantages of DLS and UV-Vis techniques: simplicity, sensitivity and selectivity to NPs, short time of measurement, and what is more, the calibration is not required. Therefore, these techniques are increasingly used for NPs characterization in many fields of science and industry [22, 23, 40]. Although the DLS technique is widely used for particles characterization, there are some problems in case of measuring samples with large-size distribution or multimodal distributions [23, 29]. If the measured colloid is monodisperse, the mean diameter of NPs can be determined using DLS technique. In case of polydisperse colloids, there is a risk that during the DLS measurement small objects can be screened by bigger ones and will not even be seen at all. By far, little has been done to determine the detection limits of DLS and UV-Vis techniques experimentally. The main aim of this study was to detect limits of widely used DLS and UV-Vis spectroscopy in characterization of polydisperse nanoparticles, colloids (Ag NPs colloids as mixture of nanoparticles with different sizes, 10 nm, and 55 nm; 10 nm and 80 nm) and to determine the possibilities to observe small objects (10 nm) in the presence of large ones (55 nm and 80 nm).

To investigate the detection limits of DLS and UV-Vis techniques colloids with the nanoparticles size about: 10, 55 and 80 nm were used. Selected colloids have nanoparticles sizes below 100 nm which is related to the definition of nanomaterials defined by the European Commission. The selected colloids give the possibility to perform high-quality and reliable measurements (the lower limit of particles size for DLS measurements is about 10 nm). The size of nanoparticles about 55 nm is in the middle of the selected limits (10 nm and 80 nm).

DLS and UV-Vis spectroscopy as well as AFM and TEM were used to characterize the monodisperse silver NPs. The colloids were intentionally mixed in appropriate volume ratio, and mixtures (polydisperse colloids) were studied using the DLS and UV-Vis techniques.

There are works that compare results from different NPs size measurement techniques [29, 41, 42], but in general the researchers do not apply to polydisperse metallic NPs with the size up to 100 nm. To the authors' knowledge, the present paper reports the results of such an investigation for the first time.

MATERIALS AND METHODS

Synthesis of Monodisperse Silver Nanoparticles

To prepare the polydisperse colloids, it was necessary to synthesize and characterize monodisperse Ag NPs colloids. Ag NPs were obtained in three different sizes: 10 nm, 55 nm, and 80 nm. For the synthesis of NPs [43], the following reagents were used: silver nitrate ($AgNO_3$, purity 99.999%, obtained from Sigma-Aldrich), and sodium citrate ($C_6H_5Na_3O_7 \cdot 2H_2O$, purity 99.0%, obtained from Sigma-Aldrich), tannic acid ($C_{76}H_{52}O_{46}$, obtained from Fluka), sodium borohydride ($NaBH_4$, purity ≥ 96%, obtained from Sigma-Aldrich). Deionized water was obtained from Deionizer Millipore Simplicity UV system.

After the synthesis, the amount of silver ions was determined using flame atomic adsorption spectrometry (AAS, Varian, Spectra 300). The amount of silver ion was less than 0.5 ppm.

Synthesis of 10 nm Ag NPs (Sample 1)

Silver NPs with an average diameter of 10 nm were prepared as follows: into 95.5 g of aqueous silver nitrate solution at the concentration of 0.017%, set on a mechanical stirrer, a mixture of sodium citrate (4.2 g, 4%) and tannic acid (0.63 g, 5%) was added. Immediately after mixing reagents, 0.7 g of solution of sodium borohydride, at the concentration of 2%, was added. After the addition of reductants, the color of the solution changed into brown. The whole mixture was vigorously stirred for 15 min. Final concentration of Ag in colloids was 100 ppm.

Synthesis of 55 nm Ag NPs (Sample 2)

Silver NPs with an average diameter of 55 nm (100 ppm) were prepared by reduction of $AgNO_3$ by sodium citrate and tannic acid. An aqueous solution of $AgNO_3$ (48.85 g, 0.016%) was heated to boil and stirred under reflux. Next the mixture of an aqueous solution of sodium citrate (0.52 g, 4%) and an aqueous solution of tannic acid (0.63 g, 5%) was added. After the addition of the reductants the color of the solution changed into yellow, which indicated the formation of the silver NPs. The solution was vigorously stirred under reflux for additional 15 min, and after that time the suspension was cooled to room temperature. Final concentration of Ag in colloids was 100 ppm.

Synthesis of 80 nm Ag NPs (Sample 3)

Silver NPs with an average diameter of 80 nm were prepared as described above, but with the different amount of silver nitrate (492.7 g, 0.048%), sodium citrate (4.2 g, 15%), and tannic acid (3.15 g, 30%). After synthesis, the solution was dissolved to final concentration of Ag in colloids which was 100 ppm.

Measurement Techniques

The resulting monodisperse NPs of silver colloids (samples 1–3) were examined using the AFM, TEM, DLS, and UV-Vis techniques.

Atomic Force Microscopy

AFM measurements of NPs were performed with the use of a commercial AFM system (Solver P47, NT-MDT), operating in air under ambient conditions. Images were typically obtained in the tapping mode using a rectangular silicon nitride cantilever (NSC 35/Si_3N_4/AlBS, MikroMasch). The composition of the 1, 2, and 3 colloids varies. In case of samples 1, 2, and 3, the nanoparticles surface stabilizer is different, so that it was necessary to use a different surface modifier for the more effective adsorption of nanoparticles on the substrate. Hence, different substrates (mica or silicon wafer) for AFM measurements were used. Moreover, the usage of a different modifier has a negligible effect on the measured size of NPs. In order to carry out AFM measurements, colloids of silver NPs were deposited on mica (sample 1) or silicon substrates (samples 2 and 3) using the procedure described in [25].

Transmission Electron Microscopy

The size and shape of Ag NPs (samples 1 and 2) were evaluated by a high resolution transmission electron microscope, JEOL JEM 1200 EX, at an accelerating voltage of 120 kV. TEM samples of the silver NPs (samples 1, 2, and 3) were prepared by placing of the product solution onto the carbon-coated copper grids, allowing the solvent to evaporate in air.

Dynamic Light Scattering

The size and size distribution of particles in the colloids were measured using a Nano ZS zetasizer system (Malvern Instruments). Measurement parameters were as follows: a laser wavelength of 633 nm (He–Ne), a scattering angle of 173° (fixed—without changing possibility), a measurement temperature of 25°C, a medium viscosity of 0.8872 mPa·s and a medium refractive index of 1.330, and material refractive index of 0.200. Before DLS measurement, the colloid was passed through a 0.2 μm polyvinylidene fluoride (PVDF) membrane. The sample was loaded into quartz microcuvette, and five measurements were performed, for which the mean result was recorded. DLS studies were carried out in two modes: general purpose mode (with normal resolution) and multiple narrow mode (with high resolution).

UV-Vis Spectroscopy

UV-Vis spectra were recorded using as a light source versatile lamps optimized for the visible-near infrared Vis-NIR (360 2000 nm), Ocean Optics, HL-2000 (tungsten halogen light sources). To collect UV-Vis spectra, the USB2000 + detector (miniature fiber optic spectrometer) was used. Before measurement, the samples were diluted one hundred times. The presented results were obtained by averaging 1000 of single measurements.

Polydisperse Sample Preparation

The prepared and well-characterized monodisperse colloids were mixed together in appropriate volume ratios. To examine the sensitivity/resolution of DLS and UV-Vis techniques, the mixtures of monodisperse colloids with the size of 10, 55, and 80 nm were measured. The list of initial colloids and polydisperse samples prepared by the mixing of appropriate monodisperse colloids is presented in Table 1.

Table 1: Summary of tested samples. The numbers indicate the percentage of the volume for silver NPs of a particular size

Sample number	Percentage volume [%]			Presented results
	10 nm AgNPs	**55 nm AgNPs**	**80 nm AgNPs**	
1	100	—	—	DLS, AFM, TEM, UV-Vis
2	—	100	—	DLS, AFM, TEM, UV-Vis
3	—	—	100	DLS, AFM, TEM, UV-Vis
4	99	1	—	DLS, UV-Vis
5	98	2	—	DLS, UV-Vis
6	97	3	—	DLS, UV-Vis
7	96	4	—	DLS, UV-Vis
8	95	5	—	DLS, UV-Vis

9	90	10	—	UV-Vis
10	80	20	—	UV-Vis
11	70	30	—	UV-Vis
12	60	40		UV-Vis
13	50	50	—	UV-Vis
14	40	60	—	UV-Vis
15	20	80	—	UV-Vis
16	99	—	1	DLS, UV-Vis
17	98	—	2	DLS, UV-Vis
18	97	—	3	DLS, UV-Vis
19	96	—	4	DLS, UV-Vis
20	95	—	5	DLS, UV-Vis
21	90	—	10	UV-Vis
22	80	—	20	UV-Vis
23	70	—	30	UV-Vis
24	60	—	40	UV-Vis
25	50	—	50	UV-Vis
26	40	—	60	UV-Vis
27	20	—	80	UV-Vis

The DLS and UV-Vis characterization was carried out for the samples 4–8 and 16–20. Additional measurements using UV-Vis spectroscopy were performed for the samples 9–15 and 21–27.

RESULTS AND DISCUSSION

Monodisperse Silver Nanoparticles

The colloids prepared as monodisperse silver NPs were measured using DLS. The size measured in DLS technique is the hydrodynamic diameter of the theoretical sphere that diffuses with the same speed as the measured nanoparticle. This size is not only connected with the metallic core of the nanoparticles (like it is in case of microscopic techniques when measuring the size of nanoparticles) but it is also influenced with all substances adsorbed on the surface of the nanoparticles (e.g., stabilizers) and the thickness of the electrical double layer (solvation shell), moving along with the particle. The thickness of the electrical double layer and its influence on the measured size of nanoparticles depend on the substances present in the colloid and on the surface of nanoparticles. As a consequence, the size measured in DLS technique is bigger in comparison with macroscopic techniques. The results obtained for the samples 1–3 are shown in Figure 1. Analyses of obtained results indicate that in the investigated colloids monodisperse NPs with size (10 ± 5) nm (Figure 1(a)), (55 ± 9) nm (Figure 1(b)), and (80 ± 20) nm (Figure 1(c)) are present. In all cases, measurements uncertainties were calculated as the standard deviation. Measurements parameters were as follows: dispersant RI = 1,330; viscosity = 0,8872 cP; temperature = 25,0°C; attenuator was set up automatically and ranged from 6 to 9. In case of all monodisperse AgNPs, the count rate was between 160 and 400 kcps. Polydispersity index (PdI) for monodisperse silver nanoparticles was smaller than 0,140.

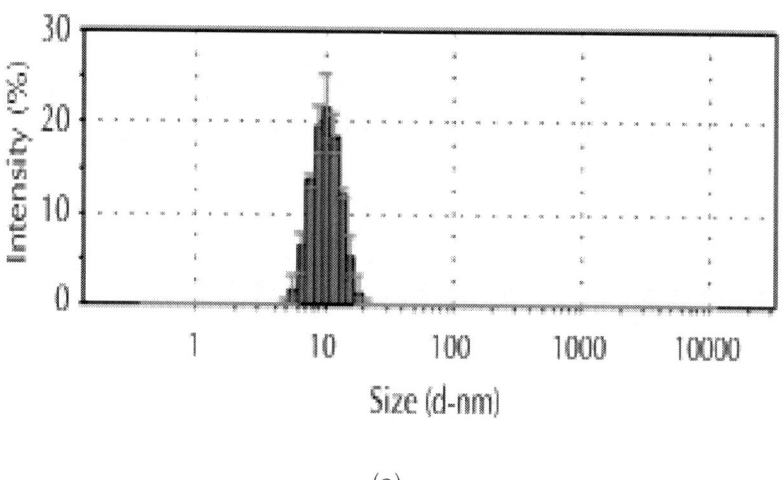

(a)

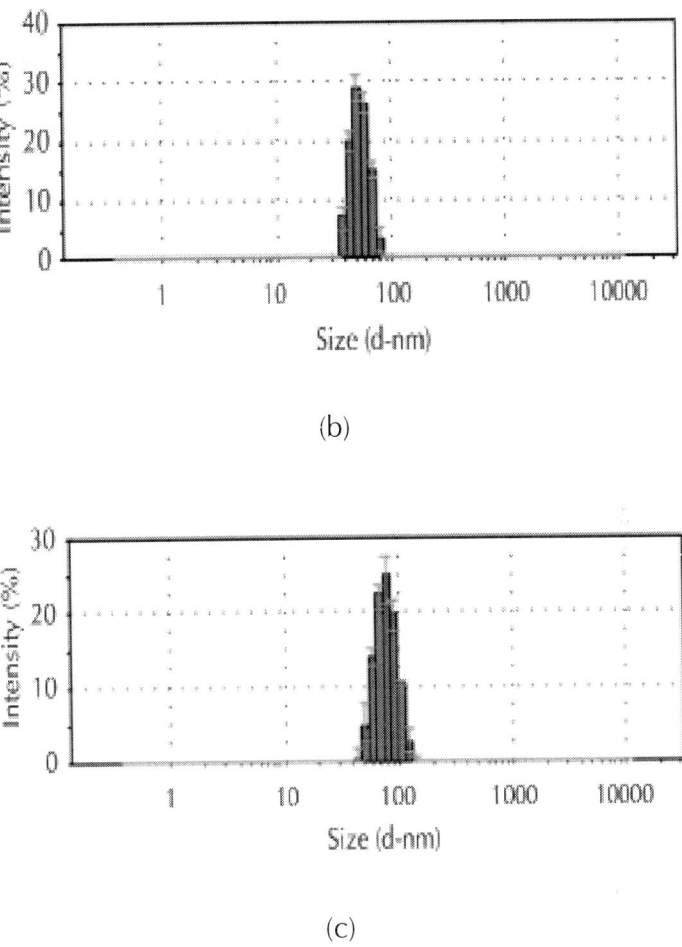

(b)

(c)

Figure 1: Size distribution of Ag NPs measured by the DLS technique for (a) sample 1, (b) sample 2, and (c) sample 3.

Figure 2 shows AFM images of Ag NPs (samples 1–3) homogeneously distributed on the different surfaces. Analysis of these images confirms that the silver NPs are monodisperse. The diameters of the obtained NPs were (8 ± 3) nm, (38 ± 6) nm and (55 ± 9) nm for the samples 1, 2, and 3, respectively. The average sizes of NPs were determined by measuring at least one hundred objects and creating a histogram. AFM images show that the investigated objects are monodisperse, without aggregates or agglomerates. The particles are characterized by a narrow size distribution around the mean value.

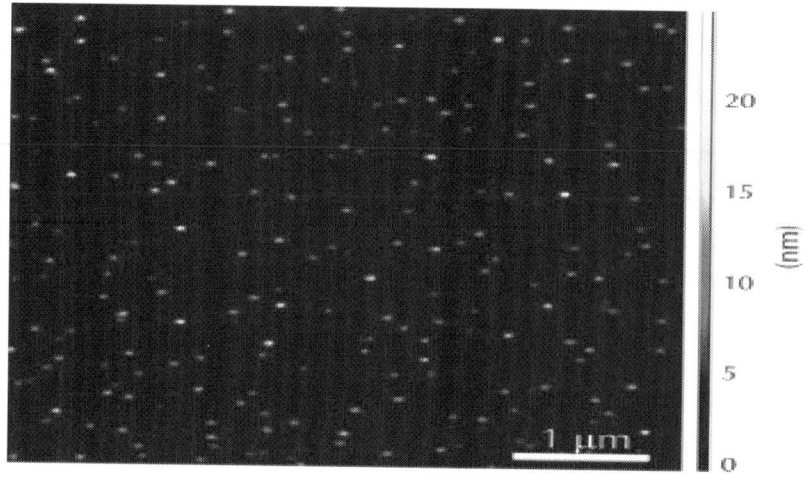

(a)

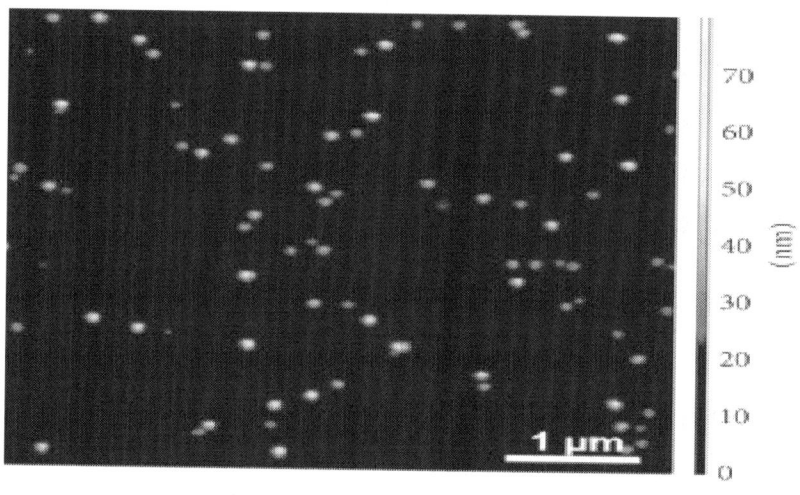

(b)

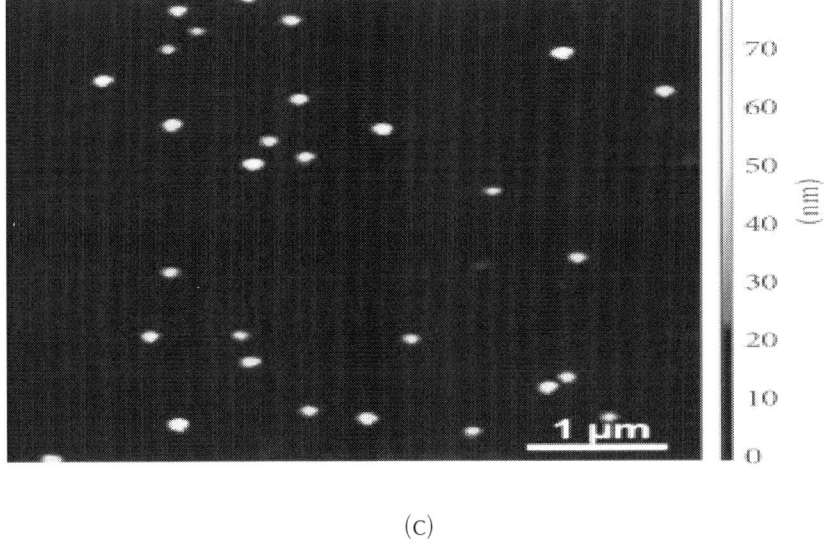

(c)

Figure 2: AFM images of monodisperse silver NPs for (a) sample 1 deposited on mica substrate, (b) sample 2 deposited on silicon substrate, and (c) sample 3 deposited on silicon substrate.

Additionally, TEM was also used to determine the average size of silver NPs. TEM images of sample 1 (Figure3(a)), sample 2 (Figure 3(b)), and sample 3 (Figure 3(c)) were recorded in order to verify the presence of monodisperse NPs. The diameters of nanoparticles were measured using Motic Plus 2.0 software. The population of analyzed nanoparticles was about 100 for AFM and TEM measurements. Next, the size distribution histogram was created, and for each data set, the mean size and standard deviation were calculated. The study allowed determining the average sizes of NPs to be (10 ± 2) nm, (41 ± 5) nm, and (61 ± 11) nm for the samples 1, 2, and 3, respectively. Agglomerated nanoparticles seen on TEM images are connected with the sample preparation for measurements. In case of sample preparation for TEM measurements, it is not possible to modify the substrate surface (copper grid with a thin carbon coating). As a consequence, it may happen that some accumulated nanoparticles can be seen. However, AFM and DLS measurements revealed that colloids (sample 1, 2, and 3) are monodisperse.

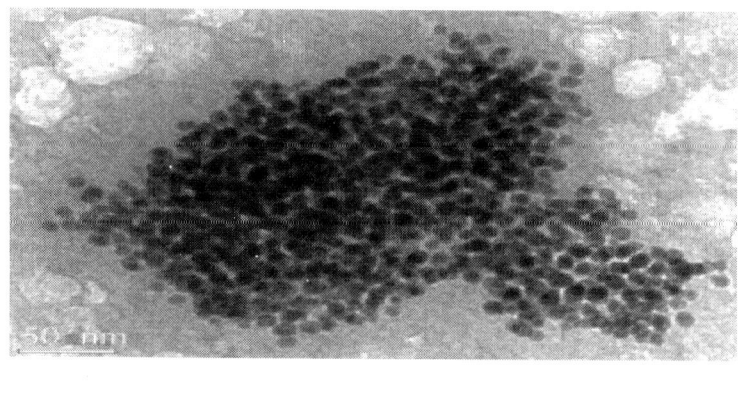

(a)

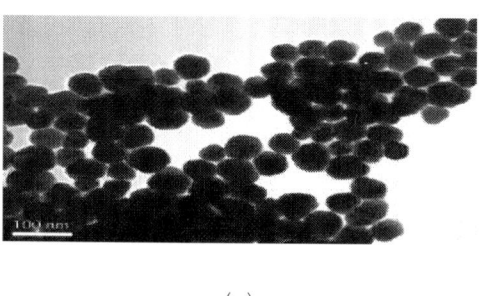

(b)

(c)

Figure 3: TEM images of monodisperse Ag NPs for (a) sample 1, (b) sample 2, and (c) sample 3.

For monodisperse colloids, measurements with the use of UV-Vis spectroscopy were also performed. The absorption peaks are located at wavelengths of 407 nm, 427 nm, and 439 nm for the samples 1, 2, and 3, respectively.

It should be noted that differences in the size of NPs were determined by the use of different techniques. This is not caused by measurement error, but in fact by the specificity of each technique. TEM and AFM measure the geometric size of the NPs deposited on the surface, so the results from these two techniques are similar. In case of the DLS technique, the hydrodynamic size is measured. This size is characterized for the ball model, which has the same diffusion coefficient as a measured NP. As a result, the size of measured NPs can differ from that determined by the AFM/TEM techniques. In general, each presented measurement method shows that only monodisperse particles are present in the investigated colloids.

Polydisperse Colloids of Silver Nanoparticles

The DLS results of the polydisperse colloids obtained by mixing monodisperse 10 nm and 55 nm Ag NPs (samples 4–8) are shown in Figure 4. Similar studies using DLS have been carried out for the polydisperse colloids obtained by mixing 10 nm with 80 nm Ag NPs (samples 16–20), and the obtained results are presented in Figure 5. It can be observed that when the volume of bigger particles increases the signal from small particles decreases. This is a consequence of the fact that the ability of a particle to scatter light is proportional to its diameter to the sixth power [41]. The peak coming from 10 nm Ag NPs disappears completely at the 95% of the content of these particles in the colloid (Figure 4(e)). In the considered case, the intensity of light scattered by larger particles (55 nm or 80 nm Ag NPs) totally conceals the signal from smaller ones (10 nm Ag NPs). And as a consequence, such a result can wrongly suggest that only monodisperse particles 55 nm (or 80 nm) in size are present in the colloid.

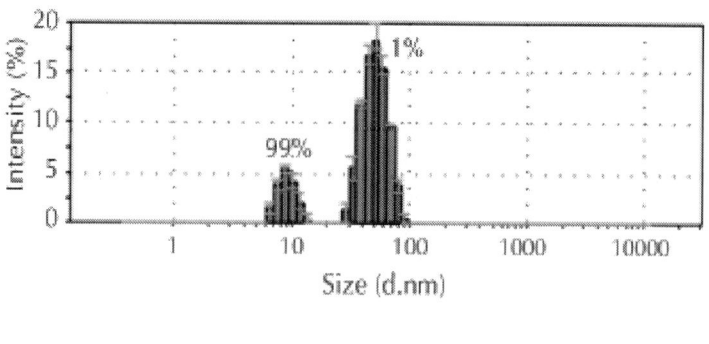

(a)

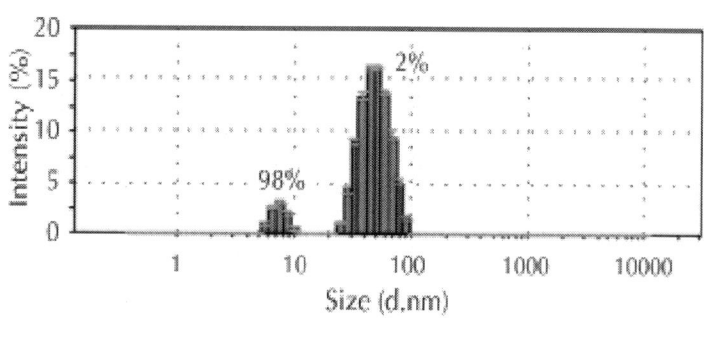

(b)

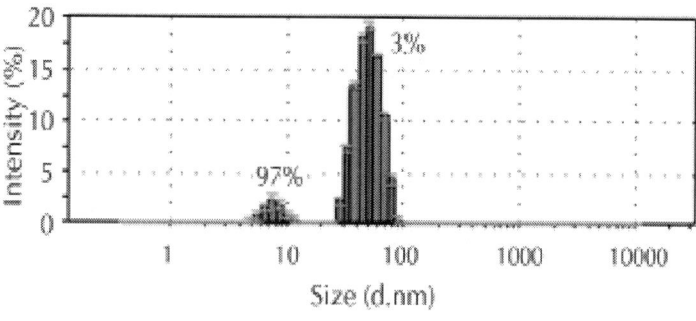

(c)

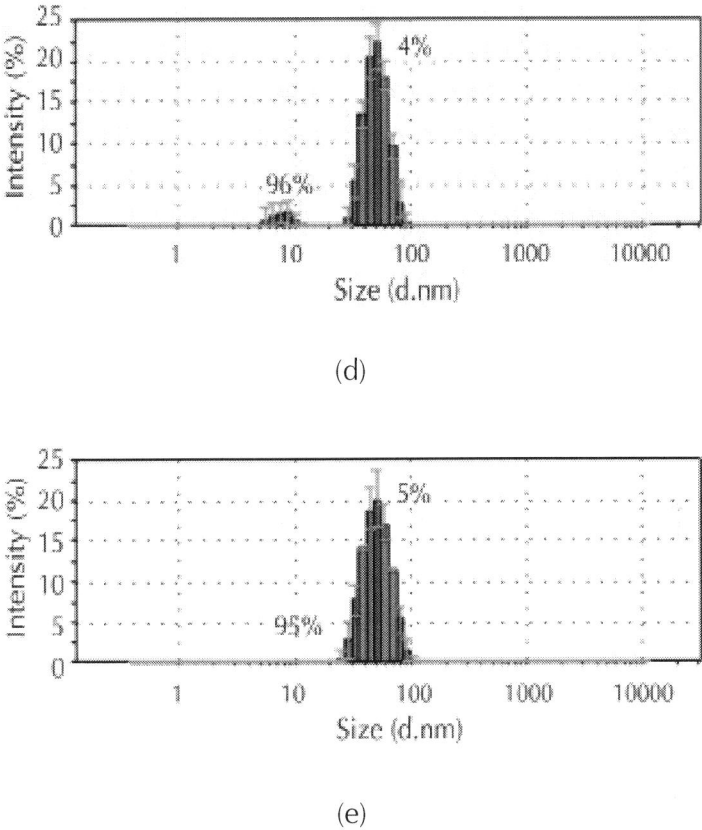

(d)

(e)

Figure 4: Size distribution of silver NPs measured by the DLS technique for mixtures of monodisperse 10 nm and 55 nm Ag NPs for (a) sample 4, (b) sample 5, (c) sample 6, (d) sample 7, and (e) sample 8. The numbers close to the histogram bars indicate the percentage of the volume for Ag NPs of a particular size in the analyzed mixture.

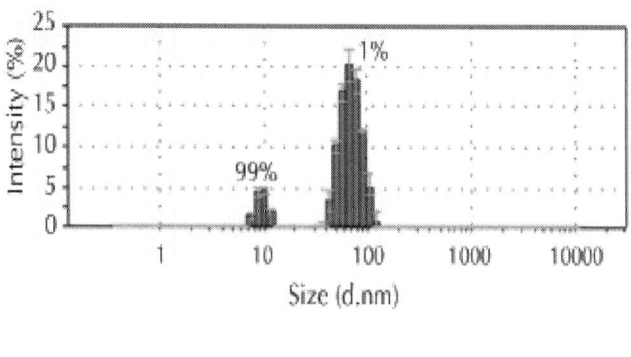

(a)

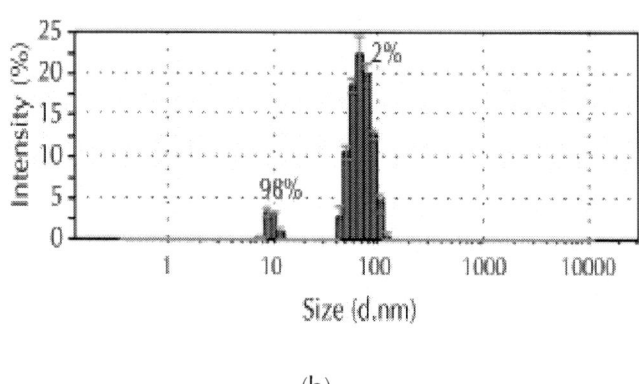

(b)

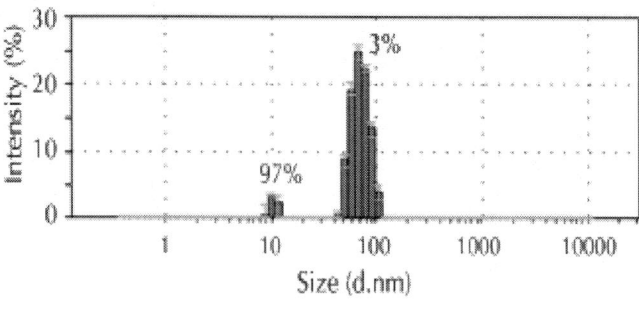

(c)

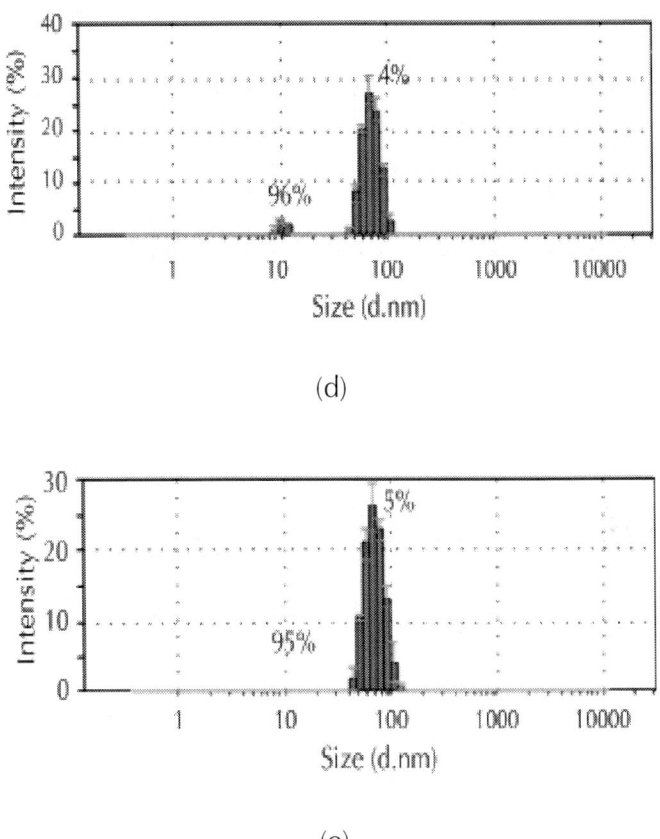

(d)

(e)

Figure 5: Size distribution of silver NPs measured by the DLS technique for mixtures of monodisperse 10 nm and 80 nm Ag NPs for (a) sample 16, (b) sample 17, (c) sample 18, (d) sample 19, and (e) sample 20. The numbers close to the histogram bars indicate the percentage of the volume for Ag NPs of a particular size in the analyzed mixture.

The characterization results of the optical properties of the polydisperse silver colloids obtained by mixing monodisperse 10 nm and 55 nm (10 nm and 80 nm) Ag NPs in the appropriate volume ratio are shown in Figure 6(a) (Figure 6(b)).

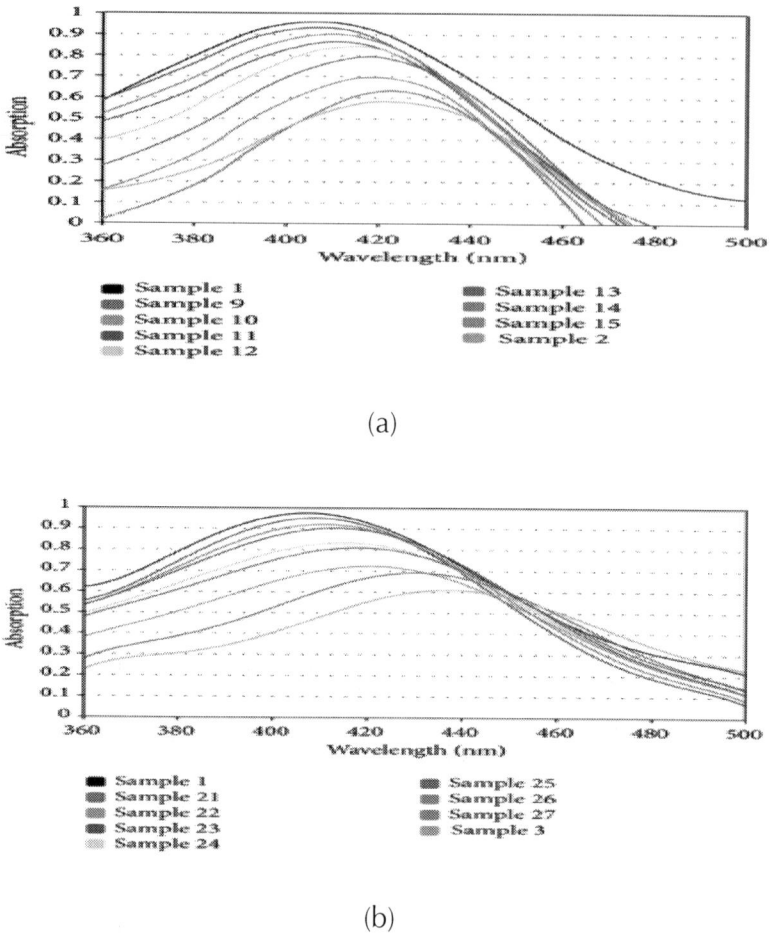

Figure 6: Evolution of the absorption spectra for mixtures of (a) 10 nm and 55 nm Ag NPs, (b) 10 nm and 80 nm Ag NPs.

It can be seen that the maximum absorption wavelength increases with the increasing of larger size Ag NPs percentage volume. Finally the maximum absorption peak is shifted to a wavelength of 427 nm (439 nm) when only particles with the size of 55 nm (80 nm) are present in the solution. Such a red-shift is characteristic for an increased NPs size [43, 44]. The position of absorption maximum of polydisperse colloid does not give information about the size of nanoparticles, because it is not possible to observe peaks separation form specific populations in these colloids.

CONCLUSIONS

Measurements of the artificially prepared polydisperse colloids were performed using the DLS and UV-Vis spectroscopy techniques. Based on the DLS results, it can be concluded that the detection of smaller NPs in the presence of several percent of bigger ones (mixed together as polydisperse colloids) seems to be in fact very difficult.

Simple calculations show that the number of particles with the size of 10 nm in the tested mixtures could be even three orders of magnitude larger than that of 55 nm particles and four orders of magnitude larger than that of 80 nm particles (under the same silver mass concentration per volume unit). However, the light scatter from bigger AgNPs is so intense that the scatter light coming from smaller AgNPs is concealed. Consequently it is not possible to detect the signal coming from 95% of smaller AgNPs in the presence of 5% bigger AuNPs. Depending on the combination of monodisperse colloids applied to prepare artificial polydisperse colloid, the detection limits of DLS and UV-Vis techniques can vary. In case of UV-Vis spectroscopy the separation of peaks for NPs of different sizes was not observed. Measurements of polydisperse colloids showed only the shift of the peak maximum compared with monodisperse samples. Hence, UV-Vis should not be used as a routine method to quantitatively examine particle size but there are a lot of researches where only UV-Vis is used to quantitatively examine particle size [45, 46].

There are other measurement techniques where using a lot more complicated equipment is possible to observe specific populations of nanoparticles in polydisperse colloids [47]. However, the main advantage of DLS and UV-Vis is that these techniques are easy to perform as well as quick and cheap. Nevertheless, the present paper clearly shows that while carrying out the synthesis of NPs, one should be very careful while interpreting the results obtained with the mentioned techniques. The obtained results are not unambiguous and an inexperienced investigator may be misled. One has to be aware that in the case of DLS or UV-Vis spectroscopy investigation of polydisperse NPs, it is necessary to make also study by much more reliable techniques such as AFM or TEM/SEM.

ACKNOWLEDGMENTS

Scientific work was supported by funds for science in 2011–2014 allocation for the cofounded international project. This work was supported by the Polish Ministry of Science and Higher Education within Research Grant no. NN 507 350435. The authors would like to thank Dr. Witold Zielinski of Technical University of Warsaw for TEM imaging.

REFERENCES

1. E. Boisselier and D. Astruc, "Gold nanoparticles in nanomedicine: preparations, imaging, diagnostics, therapies and toxicity," Chemical Society Reviews, vol. 38, no. 6, pp. 1759–1782, 2009.

2. M. V. Yezhelyev, X. Gao, Y. Xing, A. Al-Hajj, S. Nie, and R. M. O'Regan, "Emerging use of nanoparticles in diagnosis and treatment of breast cancer," The Lancet Oncology, vol. 7, no. 8, pp. 657–667, 2006.

3. Y. C. Cao, R. Jin, and C. A. Mirkin, "Nanoparticles with Raman spectroscopic fingerprints for DNA and RNA detection," Science, vol. 297, no. 5586, pp. 1536–1540, 2002.

4. J. B. Falabella, T. J. Cho, D. C. Ripple, V. A. Hackley, and M. J. Tarlov, "Characterization of gold nanoparticles modified with single-stranded DNA using analytical ultracentrifugation and dynamic light scattering," Langmuir, vol. 26, no. 15, pp. 12740–12747, 2010.

5. H. Li and L. Rothberg, "Colorimetric detection of DNA sequences based on electrostatic interactions with unmodified gold nanoparticles," Proceedings of the National Academy of Sciences of the United States of America, vol. 101, no. 39, pp. 14036–14039, 2004.

6. J.-M. Nam, C. S. Thaxton, and C. A. Mirkin, "Nanoparticle-based bio-bar codes for the ultrasensitive detection of proteins," Science, vol. 301, no. 5641, pp. 1884–1886, 2003.

7. T. A. Taton, C. A. Mirkin, and R. L. Letsinger, "Scanometric DNA array detection with nanoparticle probes," Science, vol. 289, no. 5485, pp. 1757–1760, 2000.

8. D. J. Gorth, D. M. Rand, and T. J. Webster, "Silver nanoparticle toxicity in Drosophila: size does matter,"International Journal of Nanomedicine, vol. 6, pp. 343–350, 2011.

9. M. Korani, S. M. Rezayat, K. Gilani, S. Arbabi Bidgoli, and S. Adeli, "Acute and subchronic dermal toxicity of nanosilver in guinea pig," International Journal of Nanomedicine, vol. 6, pp. 855–862, 2011.

10. H. J. Johnston, G. Hutchison, F. M. Christensen, S. Peters, S. Hankin, and V. Stone, "A review of the in vivo and in vitro toxicity of silver and gold particulates: particle attributes and biological mechanisms responsible for the observed toxicity," Critical Reviews in Toxicology, vol. 40, no. 4, pp. 328–346, 2010.

11. Y. M. Mohan, K. Lee, T. Premkumar, and K. E. Geckeler, "Hydrogel networks as nanoreactors: a novel approach to silver nanoparticles for antibacterial applications," Polymer, vol. 48, no. 1, pp. 158–164, 2007.

12. A. J. Haes, C. L. Haynes, A. D. McFarland, G. C. Schatz, R. P. Van Duyne, and S. Zou, "Plasmonic materials for surface-enhanced sensing and spectroscopy," MRS Bulletin, vol. 30, no. 5, pp. 368–375, 2005.

13. G.-N. Xiao and S.-Q. Man, "Surface-enhanced Raman scattering of methylene blue adsorbed on cap-shaped silver nanoparticles," Chemical Physics Letters, vol. 447, no. 4–6, pp. 305–309, 2007.

14. T. Yamaguchi, T. Kaya, and H. Takei, "Characterization of cap-shaped silver particles for surface-enhanced fluorescence effects," Analytical Biochemistry, vol. 364, no. 2, pp. 171–179, 2007.

15. O. V. Salata, "Applications of nanoparticles in biology and medicine," Journal of Nanobiotechnology, vol. 2, article 3, 2004.

16. J. M. Zook, V. Rastogi, R. I. MacCuspie, A. M. Keene, and J. Fagan, "Measuring agglomerate size distribution and dependence of localized surface plasmon resonance absorbance on gold nanoparticle agglomerate size using analytical ultracentrifugation," ACS Nano, vol. 5, no. 10, pp. 8070–8079, 2011.

17. S. Link and M. A. El-Sayed, "Size and temperature dependence of the plasmon absorption of colloidal gold nanoparticles," Journal of Physical Chemistry B, vol. 103, no. 21, pp. 4212–4217, 1999.

18. M. Hu, J. Chen, Z.-Y. Li et al., "Gold nanostructures: engineering their plasmonic properties for biomedical applications," Chemical Society Reviews, vol. 35, no. 11, pp. 1084–1094, 2006.

19. I. Piwonski, K. Soliwoda, A. Kisielewska, K. Kadziola, and R. Stanecka-Badura, "The effect of the surface nanostructure and composition on the antiwear properties of zirconia-titania coatings," Ceramics International, vol. 39, no. 2, pp. 1111–1123, 2013.

20. R. D. Boyd and A. Cuenat, "New analysis procedure for fast and reliable size measurement of nanoparticles from atomic force microscopy images," Journal of Nanoparticle Research, vol. 13, no. 1, pp. 105–113, 2011.

21. M. Cichomski, E. Tomaszewska, K. Ko la, W. Kozłowski, P. J. Kowalczyk, and J. Grobelny, "Study of dithiol monolayer as the interface for controlled deposition of gold nanoparticles," Materials Characterization, vol. 62, no. 3, pp. 268–274, 2011.

22. S. K. Brar and M. Verma, "Measurement of nanoparticles by light-scattering techniques," Trends in Analytical Chemistry, vol. 30, no. 1, pp. 4–17, 2011.

23. B. N. Khlebtsov and N. G. Khlebtsov, "On the measurement of gold nanoparticle sizes by the dynamic light scattering method," Colloid Journal, vol. 73, no. 1, pp. 118–127, 2011.

24. Y. Dieckmann, H. Cölfen, H. Hofmann, and A. Petri-Fink, "Particle size distribution measurements of manganese-doped ZnS nanoparticles," Analytical Chemistry, vol. 81, no. 10, pp. 3889–3895, 2009.

25. J. Grobelny, F. W. Delrio, N. Pradeep, D. Kim -I, V. A. Hackley, and R. F. Cook, "Size measurement of nanoparticles using atomic force microscopy," in Characterization of Nanoparticles Intended for Drug Delivery, S. E. McNeil, Ed., vol. 697 of Methods in Molecular Biology, pp. 71–82, Springer, 2011.

26. H. X. He, H. Zhang, Q. G. Li, T. Zhu, S. F. Y. Li, and Z. F. Liu, "Fabrication of designed architectures of Au nanoparticles on solid substrate with printed self-assembled monolayers as templates," Langmuir, vol. 16, no. 8, pp. 3846–3851, 2000.

27. S. Pethkar, M. Aslam, I. S. Mulla, P. Ganeshan, and K. Vijayamohanan, "Preparation and characterisation of silver

quantum dot superlattice using self-assembled monolayers of pentanedithiol,"Journal of Materials Chemistry, vol. 11, no. 6, pp. 1710–1714, 2001.

28. H. Jans, X. Liu, L. Austin, G. Maes, and Q. Huo, "Dynamic light scattering as a powerful tool for gold nanoparticle bioconjugation and biomolecular binding studies," Analytical Chemistry, vol. 81, no. 22, pp. 9425–9432, 2009.

29. B. G. Zanetti-Ramos, M. B. Fritzen-Garcia, C. S. de Oliveira et al., "Dynamic light scattering and atomic force microscopy techniques for size determination of polyurethane nanoparticles," Materials Science and Engineering C, vol. 29, no. 2, pp. 638–640, 2009.

30. M. Zimbone, L. Calcagno, G. Messina, P. Baeri, and G. Compagnini, "Dynamic light scattering and UV-vis spectroscopy of gold nanoparticles solution," Materials Letters, vol. 65, no. 19-20, pp. 2906–2909, 2011.

31. D. K. Bhui, H. Bar, P. Sarkar, G. P. Sahoo, S. P. De, and A. Misra, "Synthesis and UV-vis spectroscopic study of silver nanoparticles in aqueous SDS solution," Journal of Molecular Liquids, vol. 145, no. 1, pp. 33–37, 2009.

32. E. Hao, G. C. Schatz, and J. T. Hupp, "Synthesis and optical properties of anisotropic metal nanoparticles," Journal of Fluorescence, vol. 14, no. 4, pp. 331–341, 2004.

33. J. A. Creighton and D. G. Eadon, "Ultraviolet-visible absorption spectra of the colloidal metallic elements," Journal of the Chemical Society, Faraday Transactions, vol. 87, no. 24, pp. 3881–3891, 1991.

34. W. Tscharnuter, "Photon correlation spectroscopy in particle sizing," in Encyclopedia of Analytical Chemistry, R. A. Meyers, Ed., pp. 5469–5485, John Wiley & Sons, Chichester, UK, 2000.

35. D. D. Evanoff Jr. and G. Chumanov, "Synthesis and optical properties of silver nanoparticles and arrays," ChemPhysChem, vol. 6, no. 7, pp. 1221–1231, 2005.

36. D. E. Koppel, "Analysis of macromolecular polydispersity in intensity correlation spectroscopy: the method of cumulants," The Journal of Chemical Physics, vol. 57, no. 11, pp. 4814–4820, 1972.

37. B. J. Berne and R. Pecora, Dynamic Light Scattering: With Applications to Chemistry, Biology, and Physics, Dover, New York, NY, USA, 2000.

38. A. B. Leung, K. I. Suh, and R. R. Ansari, "Particle-size and velocity measurements in flowing conditions using dynamic light scattering," Applied Optics, vol. 45, no. 10, pp. 2186–2190, 2006.

39. X. Huang, P. K. Jain, I. H. El-Sayed, and M. A. El-Sayed, "Gold nanoparticles: interesting optical properties and recent applications in cancer diagnostics and therapy," Nanomedicine, vol. 2, no. 5, pp. 681–693, 2007.

40. R. Sato-Ber u, R. Redón, A. Vázquez-Olmos, and J. M. Saniger, "Silver nanoparticles synthesized by direct photoreduction of metal salts. Application in surface-enhanced Raman spectroscopy," Journal of Raman Spectroscopy, vol. 40, no. 4, pp. 376–380, 2009.

41. R. D. Boyd, S. K. Pichaimuthu, and A. Cuenat, "New approach to inter-technique comparisons for nanoparticle size measurements; using atomic force microscopy, nanoparticle tracking analysis and dynamic light scattering," Colloids and Surfaces A, vol. 387, no. 1–3, pp. 35–42, 2011.

42. C. M. Hoo, N. Starostin, P. West, and M. L. Mecartney, "A comparison of atomic force microscopy (AFM) and dynamic light scattering (DLS) methods to characterize nanoparticle size distributions,"Journal of Nanoparticle Research, vol. 10, no. 1, pp. 89–96, 2008.

43. T. Dadosh, "Synthesis of uniform silver nanoparticles with a controllable size," Materials Letters, vol. 63, no. 26, pp. 2236–2238, 2009.

44. B. J. Messinger, K. U. Von Raben, R. K. Chang, and P. W. Barber, "Local fields at the surface of noble-metal microspheres," Physical Review B, vol. 24, no. 2, pp. 649–657, 1981.

45. N. G. Khlebtsov, "Determination of size and concentration of gold nanoparticles from extinction spectra," Analytical Chemistry, vol. 80, no. 17, pp. 6620–6625, 2008.

46. N. G. Khlebtsov, L. A. Trachuk, and A. G. Mel›nikov, "The effect of the size, shape, and structure of metal nanoparticles on the

dependence of their optical properties on the refractive index of a disperse medium," Optics and Spectroscopy, vol. 98, no. 1, pp. 77–83, 2005.

47. J.-L. Fraikin, T. Teesalu, C. M. McKenney, E. Ruoslahti, and A. N. Cleland, "A high-throughput label-free nanoparticle analyser," Nature Nanotechnology, vol. 6, no. 5, pp. 308–313, 2011.

Chapter 8

Synthesis and Ultraviolet Visible Spectroscopy Studies of Chitosan Capped Gold Nanoparticles and Their Reactions with Analytes

Norfazila Mohd Sultan and Mohd Rafie Johan

Nanomaterials Engineering Research Group, Advanced Materials Research Laboratory, Department of Mechanical Engineering, University of Malaya, 50603 Lembah Pantai, Kuala Lumpur, Malaysia

ABSTRACT

Gold nanoparticles (AuNPs) had been synthesized with various molarities and weights of reducing agent, monosodium glutamate (MSG), and stabilizer chitosan, respectively. The significance of chitosan as stabilizer was distinguished through transmission electron microscopy (TEM) images and UV-Vis absorption spectra in which the interparticles distance increases whilst retaining the surface plasmon resonance (SPR) characteristics peak. The most stable AuNPs occurred for composition with the lowest (1 g) weight of chitosan. AuNPs capped with chitosan size stayed small after 1 month aging compared to bare AuNPs. The ability of chitosan capped AuNPs to uptake analyte was studied by employing amorphous carbon nanotubes (α-CNT), copper oxide (Cu_2O), and zinc sulphate ($ZnSO_4$) as the target material. The absorption spectra showed dramatic intensity increased and red shifted once the analyte was added to the chitosan capped AuNPs.

INTRODUCTION

Gold nanoparticles (AuNPs) have glowing prospects in many applications due to their distinctive optical, electronic, and electrical properties [1]. AuNPs display intense colours when induced by incident light field. These were contributed by collective electron oscillation that gives intensification to the surface plasmon resonance (SPR) absorption.

There are various techniques to produce AuNPs such as microemulsion, reversed micelles, seeding growth, sonochemistry, photochemistry, radiolysis, and direct chemical reduction [2–4]. The most simple, economical, and powerful synthesis is the direct chemical reduction method. In the case of AuNPs, chemical reduction routes generate zerovalent gold colloids from gold precursors [5].

The invention of zerovalent gold colloids was pioneered by Turkevich et al. [6] and later refined by Frens [7] in which the ratio of gold precursors to citrate was varied. Brust-Schiffin [8] commenced the synthesis of AuNPs in organic solvents which involves a phase transfer agent such as toluene. The above conventional methods had many shortcomings which contributed to explorations of other reducing agents and alternative routes. The synthesis of AuNPs through Turkevich et al. approaches takes a longer time (1 hr) for gold salt reduction. While the use of organic solvents in Brust-Schiffin method leaves them inapt for detecting biomolecules and biological surfaces like proteins and saccharides [9], various chemicals had been exploited as reducing agent to produce zerovalent gold colloids such as amino acid derivatives like lysine and valine but without success. However, other acidic amino acid derivatives such as aspartic acid [10] and monosodium glutamate (MSG) [11] are competent in reducing gold salt (Figure 1). Sugunan and Dutta [11] produced AuNPs by emphasizing on lower molar ratio of MSG.

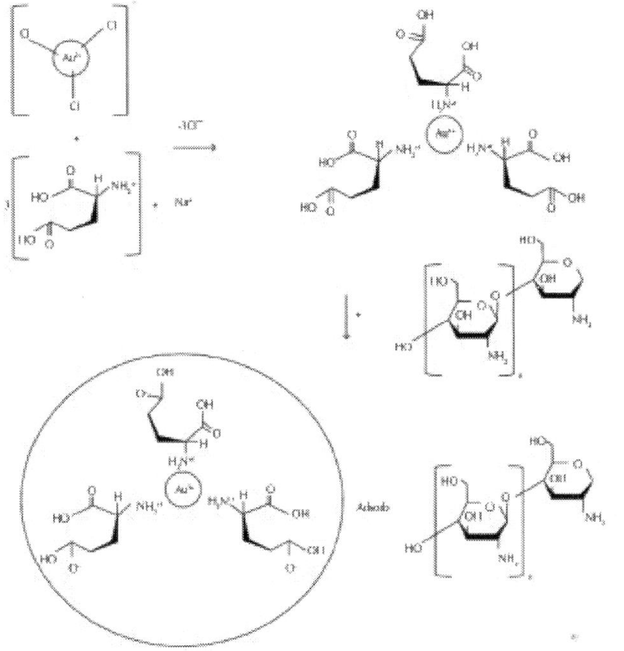

Figure 1: Schematic diagram of gold precursor reduction by MSG and capping the gold particle surfaces with chitosan.

AuNPs have compelling tendency to flocculate due to their van der walls forces. However, the agglomeration can be hindered by introducing a repulsive force between the particles. In this light, the use of stabilizer as a repulsive force came into the picture. The use of chitosan as a stabilizer was reported elsewhere [12]. Chitosan contributed the steric hindrance to stabilize the nanoparticles as shown in Figure 1. The amino group presence in its polycationic structure activates steric hindrance, thus ensuring strong stability over long durations [12]. For most biological applications, chitosan possesses many attractive functional groups such as biotin [13], aptamers [14, 15], concanavalin (con-A) [16], and bovine serum albumin [17, 18]. However, proteins have a downside as they are expensive although they were widely exploited and offer excellent characteristics. Remarkably, chitosan possesses similar ability as proteins and manipulations of its properties have not been fully extended for numerous applications. Chitosan is accessible for cross-linking through its boundless amino group and its cationic features allowing the ionic cross-linking to take place with multivalent elements. The most promising features of chitosan are its solubility in aqueous acidic solutions [19]. The description of chitosan agrees with the aims of the research to manufacture a readily biocompatible and nontoxic chitosan capped gold nanoparticles.

Sugunan et al. [9] had employed chitosan as stabilizer for silver nanoparticles for heavy metal ion sensor, yet the thermodynamically proficiency of chitosan had not been investigated. Moreover, the performance of chitosan adsorption on the surface of AuNPs has not been studied. In this paper, we report the synthesis of AuNPs and its stabilization mechanism using chitosan. By exploiting the chemistry of amine and chitosan, we have shown that the AuNPs can be prepared in water by complexation of high molar ratio glutamic acid molecules with gold precursors stabilized by the adsorption of chitosan on the surface of AuNPs. Preparation of AuNPs capped with chitosan was carried out in a single-pot process and the resulting particles were thoroughly characterized. The stability of chitosan was furthered studied and discussed.

EXPERIMENTAL SECTION

Materials

Gold (III) chloride ($AuCl_3$), acetic acid, and monosodium glutamate (MSG) (99% Na salt of L-glutamic acid) were purchased from Acros Organics. Meanwhile, chitosan (industrial grade) was purchased from Easter Holding Co. Ltd. with deacetylation degree of 80%. All chemicals were used without further purifications and all the solutions were prepared with distilled water.

Preparation of AuNPs

2 mL of 5 mM AuCl$_3$ solution (0.1517 g in 100 mL water) was stirred and heated to 100°C. Then, 3 mL of 50 mM Na salt of L-glutamic acid solution (MSG) (0.9357 g in 100 mL water) is quickly poured into the gold solution. The solution was stirred continuously until the colour changed from pale red to intense red. The steps were repeated with 100, 150, 200, 250, and 300 mM of MSG. Another set of samples was prepared for observing the aging behaviour of the AuNPs. The samples were left in ambience temperature for a month.

Preparation of Chitosan Capped AuNPs

The chitosan solution was prepared by mixing the said amount of blended chitosan powder as purchased with distilled water and adequate amount of acetic acid. The solution was stirred at room temperature until the chitosan powder had completely dissolved in the water. 990 μL of chitosan solution (1 g of chitosan in mixture of 100 mL water and 150 μL acetic acid) was then added to the as-synthesized 50 mM of MSG reduced AuNPs. A visible change of colour occurred immediately. The heating was discontinued to allow the solution to reach the ambient temperature. The steps were repeated with different concentrations of MSG (100, 150, 200, 250, and 300 mM). Another set of samples was prepared for observing the aging behaviour of the chitosan capped AuNPs. The samples were left in ambience temperature for a month.

Preparation of Amorphous Carbon Nanotubes (α-CNTs) Chitosan Capped AuNPs

The synthesis procedure of α-CNTs is followed by Tan et al. [20]. The procedure was instigated with mixture of 8 mL of ethyl alcohol (90%), 4.2 g of NaBH$_4$ (99.99%), and 15 mL of 1 M NaOH in a 25 mL flask. The solution was further stirred for the next 45 minutes before being transferred to a Parr reactor with capacity of 200 mL. The reactor was heated inside a furnace up to 200°C and held for 2 hours under scaled condition. The Parr reactor was allowed to cool to ambient temperature and the precipitate was washed thoroughly with alcohol and deionised water. The precipitate was then dried in the vacuum oven. α-CNTs were added to the optimum condition of chitosan capped AuNPs solution (1 g of chitosan powder and 100 mM of MSG).

Preparation of Copper Oxide-Chitosan Capped AuNPs

0.005, 0.01, 0.05, 0.1, and 0.5 g of purchased copper oxide powder were added to optimum condition of chitosan capped AuNPs solution (1 g of chitosan powder and 100 mM of MSG).

Preparation of Zinc Sulphate-Chitosan Capped AuNPs

0.005, 0.01, 0.05, 0.1, and 0.5 g of purchased zinc sulphate powder were added to optimum condition of chitosan capped AuNPs solution (1 g of chitosan powder and 100 mM of MSG).

Characterizations of AuNPs

Transmission electron microscope (Libra 120 TEM using accelerating voltage of 400 kV) was employed to assess the particles size and distribution of the particles. The optical properties of gold dispersions were investigated by UV-Vis spectrophotometer using UVIKON 923 UV-Vis spectrophotometer.

RESULTS AND DISCUSSION

TEM Analysis

Figures 2, 3, and 4 show the TEM images for AuNPs prepared at different concentrations of MSG. The particles are nearly spherical with high dispersibility. The average size of particles for 100, 200, and 300 mM MSG is 18, 15, and 9 nm, respectively. It is clearly shown that high molar of MSG produces smaller particle size.

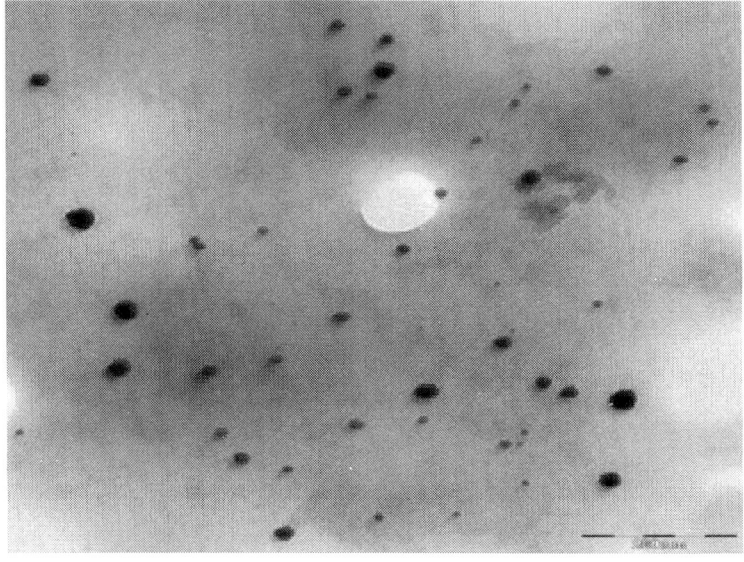

Figure 2: TEM image of AuNPs reduced with 100 mM of MSG.

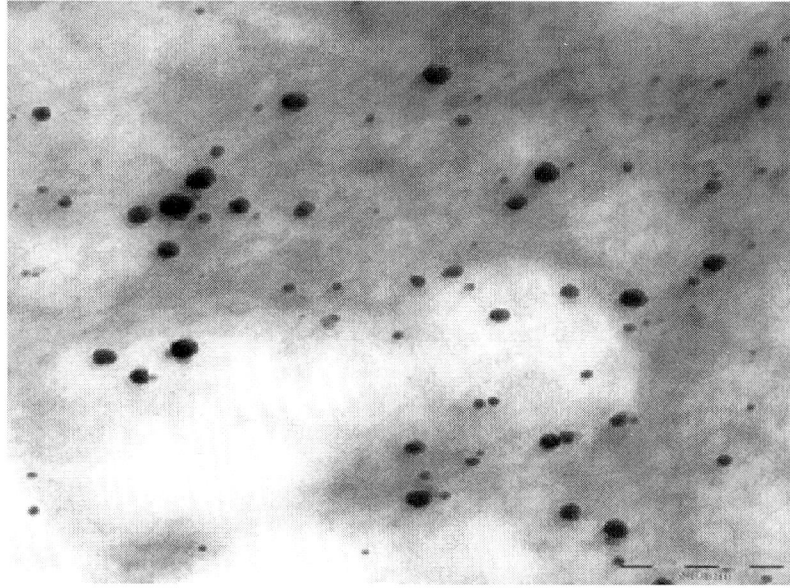

Figure 3: TEM image of AuNPs reduced with 200 mM of MSG.

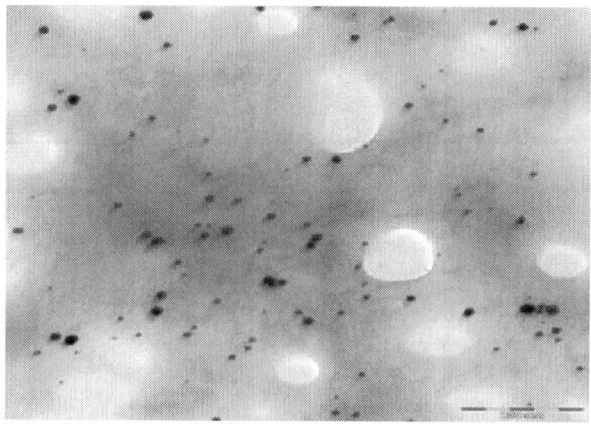

Figure 4: TEM image of AuNPs reduced with 300 mM of MSG.

The role of chitosan in steric mechanism has been verified by the TEM image shown in Figure 5. The chitosan which resembles a spider web infused a repelling force between the AuNPs separating them apart unlike the bare AuNPs (Figures 2–4). The average interparticles distance increases to 96 nm due to wrapping of chitosan around the AuNPs.

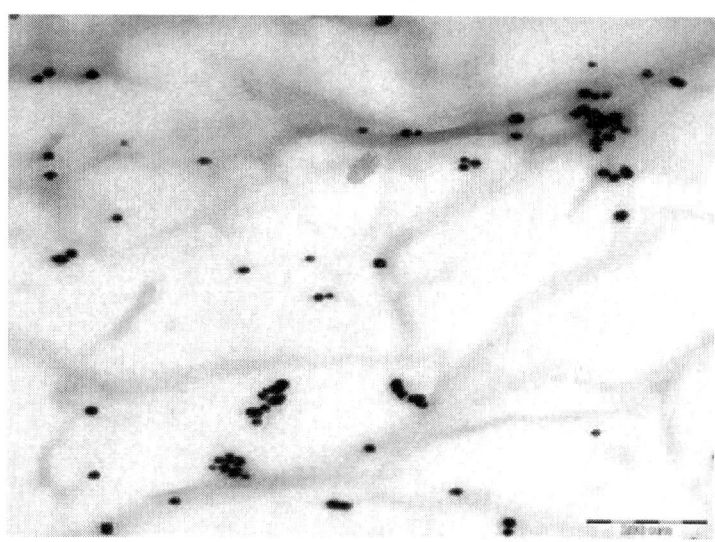

Figure 5: TEM image of chitosan capped AuNPs.

UV-Vis Spectroscopy Analysis

Effect of Concentration of Reducing Agent MSG

Figure 6 shows the absorption spectra of AuNPs at various concentrations of MSG. The surface plasmon resonance (SPR) peaks are shifted to the smaller wavelengths indicating the reduction in particle sizes. This result is in good agreement with the TEM images in Figures 2–4. The symmetrical shape of the absorption spectra indicates that sample has a narrow particle size distribution.

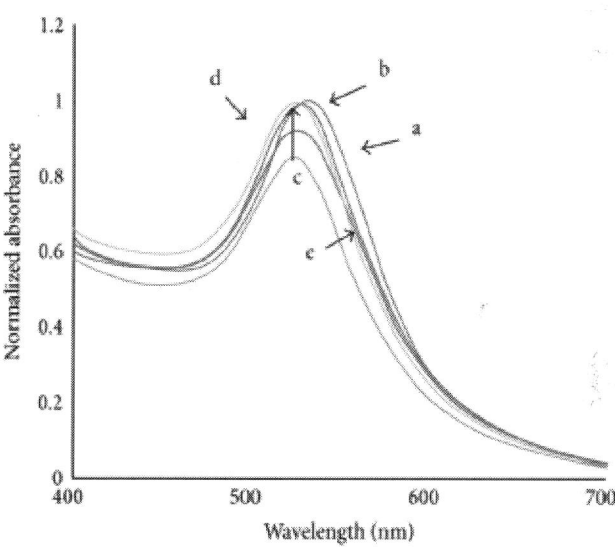

Figure 6: Absorption spectra of AuNPs with various concentrations of reducing agent MSG: (a) 100 mM; (b) 150 mM; (c) 200 mM; (d) 250 mM; (e) 300 mM.

Effect of Chitosan as Stabilizer

Figure 7 shows the absorption spectra of bare and increased weight of chitosan capped AuNPs at the optimum concentration of reducing agent

MSG (100 mM). The absorbance of chitosan capped AuNPs is higher than bare AuNPs. The SPR peak is shifted to the longer wavelength for chitosan capped AuNPs. This red shifted trend is continued for samples with increasing weight of chitosan. The absorbance for chitosan capped AuNPs is slightly increased with the increase of chitosan weight. The attachment of chitosan on the surface of AuNPs affected their optical properties.

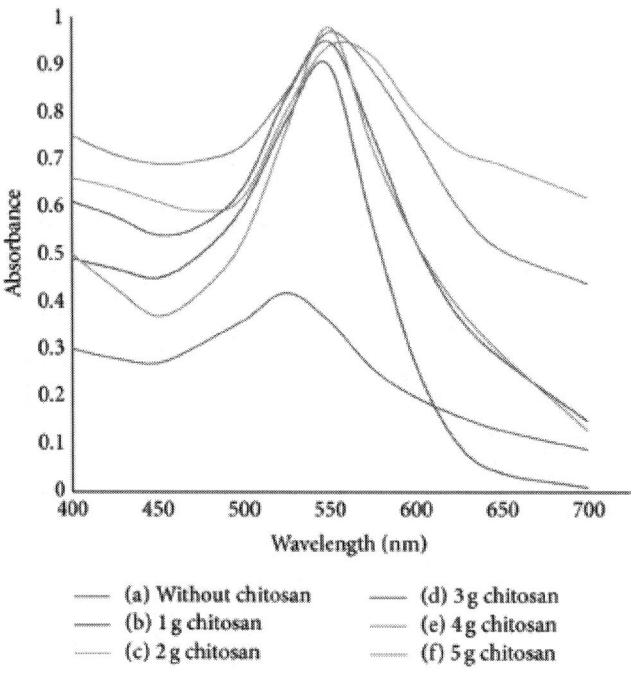

Figure 7: Absorption spectra of bare and increased weight of chitosan capped AuNPs at the optimum concentration of reducing agent MSG: (a) 0 g chitosan; (b) 1 g chitosan; (c) 2 g chitosan; (d) 3 g chitosan; (e) 4 g chitosan; (f) 5 g chitosan.

Effect of Aging

Figure 8 shows the absorption spectra of AuNPs with various concentrations of MSG after 1 month aging time. The SPR peaks are shifted to longer wavelength compared to their counterparts in Figure

6. The same goes for their FWHM values which show more broadened peak after 1 month ageing time. This indicates that the AuNPs size and their particle size distribution are increased after aging.

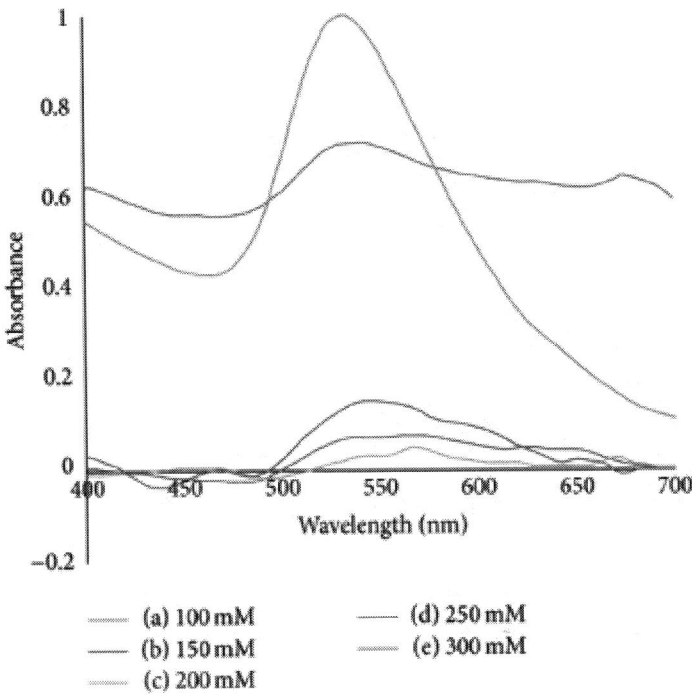

Figure 8: Absorption spectra of AuNPs with various concentrations of reducing agent MSG after 1 month aging time: (a) 100 mM; (b) 150 mM; (c) 200 mM; (d) 250 mM; (e) 300 mM.

Figure 9 shows the absorption spectra of bare and chitosan capped AuNPs for various concentrations of MSG after 1 month ageing time. The SPR peaks are shifted to the smaller wavelength compared to the bare AuNPs (Figure 8). The same goes for their FWHM values which are smaller than the FWHM values of the bare AuNPs. These absorption spectra have highlighted the role of chitosan adsorption on the AuNPs surfaces, in which their particle size stays small even after 1 month aging time. Chitosan preserves stability and hinders agglomeration of AuNPs. The FWHM and other experimental results are listed in Table 1.

Table 1: Experimental values of SPR peaks for bare and chitosan capped AuNPs with full width half maximum (FWHM) after 1 month ageing time

Sample	SPR I_{max} (nm)		FHWM (nm)	
Concentration of MSG (mM)	AuNPs	Chitosan capped AuNPs	AuNPs	Chitosan capped AuNPs
100	532	531	51	60
150	544	537	72	52
200	554	539	52	51
250	544	527	96	46
300	534	526	70	57

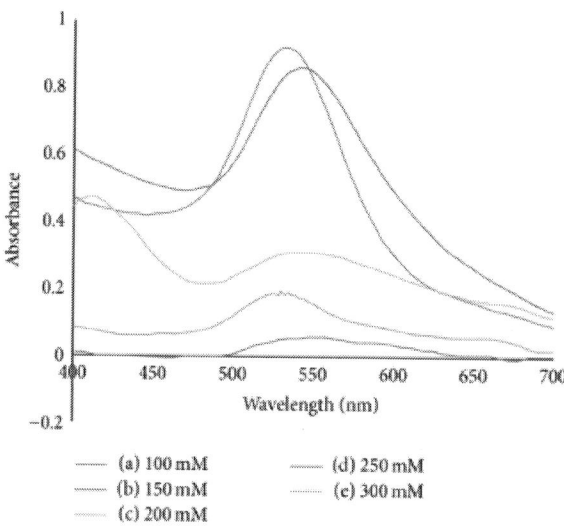

Figure 9: Absorption spectra of chitosan capped AuNPs with various concentrations of reducing agent MSG after 1 month aging time: (a) 100 mM; (b) 150 mM; (c) 200 mM; (d) 250 mM; (e) 300 mM.

Effect of Amorphous Carbon Nanotube (α-CNTs)s, Oxides, and Sulphate to Chitosan Capped AuNPs

Figure 10 shows the SPR peaks of chitosan capped AuNPs adjourned at 549 nm for three different weights of α-CNTs. The SPR peak intensity rises as the weight of α-CNTs increases. This phenomenon can be explained with regard to the fact that AuNPs are very sensitive in the weight change of α-CNTs upon exposure. High surface ratios of AuNPs contribute to the sensitivities and make them more reactive and able to uptake the analyte.

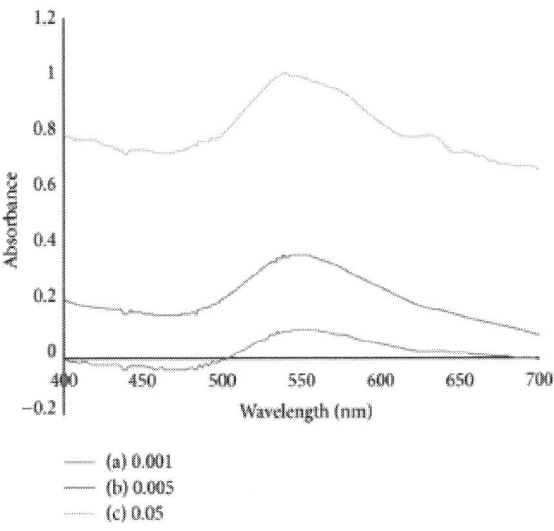

Figure 10: Absorption spectra of chitosan capped AuNPs mixed with different weights of α-CNTs: (a) 0.001 g; (b) 0.005 g; (c) 0.05 g.

Figure 11 shows the absorption spectra of chitosan capped AuNPs mixed with different weights of copper oxide. The SPR peaks are shifted from 521 to 577 nm as the weight of copper oxide increases. This clearly shows the complexation of chitosan capped AuNPs towards the addition of copper oxide. The enlargement of the particles uptake can be underlined as the peak intensity also shows dramatic increase as the weight of copper oxide increases.

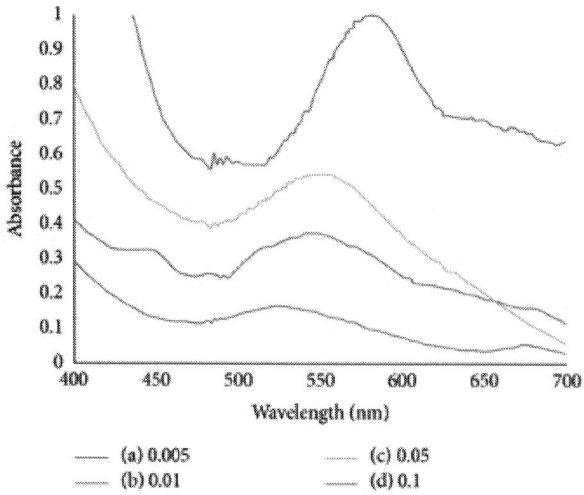

Figure 11: Absorption spectra of chitosan capped AuNPs mixed with different weights of copper oxide: (a) 0.005 g; (b) 0.01 g; (c) 0.05 g; (d) 0.1 g.

Figure 12 shows the absorption spectra of chitosan capped AuNPs at different weights of zinc sulphate. The spectra also successfully show analyte particles entrapment by chitosan capped AuNPs. The SPR peaks intensities are increased as the weight of zinc sulphate added to the chitosan capped AuNPs increases.

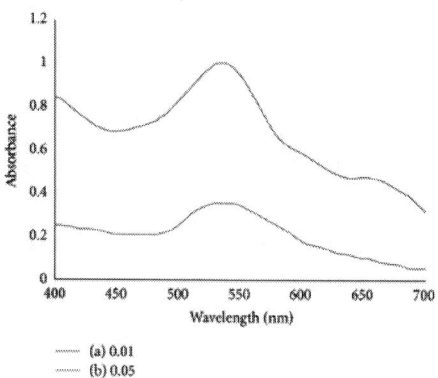

Figure 12: Absorption spectra of chitosan capped AuNPs mixed at different weights of zinc sulphate (a) 0.01 g; (b) 0.05 g.

CONCLUSIONS

We have successfully synthesized chitosan capped AuNPs via chemical reduction technique. We have also revealed that AuNPs revolutionize their dimension and optical behaviour through variation of parameters such as concentration of reducing agent, weight of stabilizer, and aging time. Chitosan concentration plays an important role in imparting extra hindrance strength. The particles stability was contributed by chitosan even after 1 month of aging. The chitosan capped AuNPs were able to uptake analyte such as α-CNTs, copper oxide, and zinc sulphate.

ACKNOWLEDGMENTS

The authors would like to express their greatest gratitude to the Ministry of Higher Education of Malaysia and University of Malaya for funding through UM/MOHE-HIR Research Grant (UM.C/HIR/MOHE/ENG/12). Special thanks are due to Mr. Low Chien Chong for providing the TEM images.

REFERENCES

1. D. A. Handley, Colloidal Gold: Principles, Methods, and Applications, Academic Press, New York, NY, USA, 1989, edited by M. A. Hayat.

2. C. Roos, M. Schmidt, J. Ebenhoch, F. Baumann, B. Deubzer, and J. Weis, "Design and synthesis of molecular reactors for the preparation of topologically trapped gold clusters,"Advanced Materials, vol. 11, no. 9, pp. 761–766, 1999. · ·

3. R. G. Freeman, M. B. Hommer, K. C. Grabar, M. A. Jackson, and M. J. Natan, "Ag-clad Au nanoparticles: novel aggregation, optical, and surface-enhanced Raman scattering properties," Journal of Physical Chemistry, vol. 100, no. 2, pp. 718–724, 1998.

4. L. Cao, P. Diao, L. Tong, T. Zhu, and Z. Liu, "Surface-enhanced Raman scattering of p-aminothiophenol on a Au(core)/Cu(shell) nanoparticle assembly," ChemPhysChem, vol. 6, no. 5, pp. 913–918, 2005. · ·

5.   N. Toshima and T. Yonezawa, "Bimetallic nanoparticles: novel materials for chemical and physical applications," New Journal of Chemistry, vol. 22, no. 11, pp. 1179–1201, 1998. · ·

6.   J. Turkevich, P. C. Stevenson, and J. Hillier, "A study of the nucleation and growth processes in the synthesis of colloidal gold," Discussions of the Faraday Society, vol. 11, pp. 55–75, 1951. · ·

7.   J. Frens, "Controlled nucleation for the regulation of the particle size in monodisperse gold suspensions," Nature Physical Science, vol. 241, pp. 20–22, 1973. ·

8.   M.-C. Daniel and D. Astruc, "Gold nanoparticles: assembly, supramolecular chemistry, quantum-size-related properties, and applications toward biology, catalysis, and nanotechnology," Chemical Reviews, vol. 104, no. 1, pp. 293–346, 2004. · ·

9.   A. Sugunan, C. Thanachayanont, J. Dutta, and J. G. Hilborn, "Heavy-metal ion sensors using chitosan-capped gold nanoparticles," Science and Technology of Advanced Materials, vol. 6, no. 3-4, pp. 335–340, 2005. · ·

10.   S. Mandal, P. R. Selvakannan, S. Phadtare, R. Pasricha, and M. Sastry, "Synthesis of a stable gold hydrosol by the reduction of chloroaurate ions by the amino acid, aspartic acid," Journal of Chemical Sciences, vol. 114, no. 5, pp. 513–520, 2002.

11.   A. Sugunan and J. Dutta, "Novel synthesis of gold nanoparticles in aqueous media," inProceedings of the Material Research Society Fall Conference, MRS, Boston, Mass, USA, 1999.

12.   H. C. Warad, S. C. Ghosh, C. Thanachayanont, and J. Dutta, "Highly luminescent manganese doped ZnS quantum dots for biological labeling," in Proceedings of International Conference on Smart Materials, Smart/Intelligent Materials and Nanotechnology (SmartMat ‹04), pp. 203–207, Chiang Mai, Thailand, 2004.

13.   N. Nath and A. Chilkoti, "Label-free biosensing by surface plasmon resonance of nanoparticles on glass: optimization of nanoparticle size," Analytical Chemistry, vol. 76, no. 18, pp. 5370–5378, 2004. · ·

14.   V. Pavlov, Y. Xiao, B. Shlyahovsky, and I. Willner, "Aptamer-functionalized Au nanoparticles for the amplified optical detection of thrombin," Journal of the American Chemical

Society, vol. 126, no. 38, pp. 11768–11769, 2004. · ·

15. C. Huang, Y. Huang, Z. Cao, W. Tan, and H. Chang, "Aptamer-modified gold nanoparticles for colorimetric determination of platelet-derived growth factors and their receptors,"Analytical Chemistry, vol. 77, no. 17, pp. 5735–5741, 2005. · ·

16. D. C. Hone, A. H. Haines, and D. A. Russell, "Rapid, quantitative colorimetric detection of a lectin using mannose-stabilized gold nanoparticles," Langmuir, vol. 19, no. 17, pp. 7141–7144, 2003. · ·

17. K. Fujiwara, H. Watarai, H. Itoh, E. Nakahama, and N. Ogawa, "Measurement of antibody binding to protein immobilized on gold nanoparticles by localized surface plasmon spectroscopy," Analytical and Bioanalytical Chemistry, vol. 386, no. 3, pp. 639–644, 2006. · ·

18. F. Frederix, J. Friedt, K. Choi et al., "Biosensing based on light absorption of nanoscaled gold and silver particles," Analytical Chemistry, vol. 75, no. 24, pp. 6894–6900, 2003. · ·

19. F. L. Mi, S. S. Shyu, C. Y. Kuan, S. T. Lee, K. T. Lu, and S. F. Ajng, "Chitosan-polyelectrolyte complexation for the preparation of gel beads and controlled release of anticancer drug. I. Effect of phosphorus polyelectrolyte complex and enzymatic hydrolysis of polymer," Journal of Applied Polymer Science, vol. 74, pp. 1868–1879, 1999.

20. K. H. Tan, R. Ahmad, B. F. Leo, M. C. Yew, B. C. Ang, and M. R. Johan, "Physico-chemical studies of amorphous carbon nanotubes synthesized at low temperature," Materials Research Bulletin, vol. 47, no. 8, pp. 1849–1854, 2012. · ·

Detection of Triphenylmethane Drugs in Fish Muscle by Surface-Enhanced Raman Spectroscopy Coupled with Au-Ag Core-Shell Nanoparticles

Lu Pei[1], Yiqun Huang[1, 2], Chunying Li[1], Yuanyuan Zhang[1], Barbara A. Rasco[2], and Keqiang Lai[1]

[1]College of Food Science and Technology, Shanghai Ocean University, No. 999 Hucheng Huan Road, LinGang New City, Shanghai 201306, China

[2]School of Food Science, Washington State University, Pullman, WA 99164, USA

ABSTRACT

Silver-coated gold bimetallic nanoparticles were synthesized and used as substrates for surface-enhanced Raman spectroscopy (SERS) in detecting prohibited triphenylmethane drugs (including crystal violet and malachite green) in fish muscle. The optical properties and physical properties of bimetallic nanospheres were characterized by UV-Vis spectroscopy and transmission electron microscopy. The optimal nanospheres selected had relatively uniform size (diameter: 33 ± 3 nm) with a silver layer coated on the surface of gold seed (diameter: 18 ± 2 nm). For both crystal violet and malachite green, characteristic SERS spectral features could be identified at concentration as low as 0.1 µg/L with these bimetallic nanospheres. Crystal violet and malachite green residues in fish muscle could also be detected at levels as low as 0.1 ng/g, which could meet the most restricted regulatory requirements for the limit of detection in terms of analytical methods for crystal violet or malachite green in fish muscle. This study provides a basis for applying SERS technology with bimetallic nanoparticles to the identification of trace amounts of prohibited substances in aquatic food products, and the methodology could be extended to analyses of other hazardous chemicals in complex food matrices like vegetables and meats.

INTRODUCTION

Surface-enhanced Raman spectroscopy or surface-enhanced Raman scattering (SERS) utilizes the tremendous enhancement effect of Raman scattering signals through adsorbing analytes onto the roughened surfaces of gold, silver, and other metallic materials [1, 2]. Chemical enhancement and electromagnetic enhancement are two accepted enhancement mechanisms, which contributed to the charge transfers between the adsorbed analyte molecule and metal substrate and the large local field enhancement in the vicinity of metal surfaces excited by the laser [3, 4]. SERS has shown great potential for analyses of trace amounts of chemicals at as low as single molecule level [5–7]. In recent years, SERS technology has been increasingly exploited in applications in various fields, such as material, physics, medicine, and food science [8, 9]. It is well known that SERS effect is quite sensitive to the changes in substrate materials and surface morphology. One of the key factors for successful applications of SERS technology is preparation of highly active, stable, and reproducible SERS substrates [2, 10]. Au-Ag core-shell bimetallic nanoparticles (NPs) were synthesized by coating silver on the surface of gold seeds through chemical reduction. With appropriate particle sizes and thickness of silver shells, these bimetallic nanospheres could overcome problems, such as instability commonly tied to silver NPs and relatively low enhancement effects tied to gold NPs, and exhibit high enhancement effects similar to silver NPs with the advantages of high degree of homogeneity as gold NPs [8, 11–13]. However, similar to other SERS substrates with great potential as analytical tools, the application aspect of bimetallic nanospheres as SERS substrates is left far behind the theory aspect of SERS related research and has become the bottleneck of the field.

Crystal violet (CV, 4-[bis(4-dimethylaminophenyl) methylidene] cyclohexa-2,5-dien-1-ylidene]-dimethylazanium chloride) and malachite green (MG, 4-[(4-dimethylaminophenyl)-phenyl-methyl]-N,N-dimethyl-aniline), both triphenylmethane dyes, are remarkably effective against fungal infections and parasitosis in fish and had been worldwide used as biocide and fungicide in aquaculture for decades [14]. Since mid-1980s, efforts have been made to reduce the use of both CV and MG due to their links to genotoxicity and carcinogenicity [15]. Although the use of these drugs has been banned in the United States, the European Union, China, and many other countries [16], CV and MG are still used illegally in some parts of the world due to their low cost and high efficiency, resulting in frequent occurrence of aquatic products with safety issues and consequently leading to import bans and product recalls. Based upon the Rapid Alert System for Food and Feed online searchable database of the European Union, a total of 131 cases of various fish products (such as catfish, trout, tilapia, salmon, king prawn, and caviar) contaminated with CV or MG were reported from 2003 to 2012 [17]. Food safety incidents like these have brought huge barriers to import and export trades of global aquatic products, so it is particularly important to strengthen the monitoring system for CV and MG residues in aquatic products. High performance liquid chromatography (HPLC) is the most commonly used method for analyzing CV and MG drug residues in fish, but it has disadvantages such as high cost, time consuming, and too complicated sample preparation [18]. In addition, with the increasing tightened policy towards CV and MG, the sensitivity of HPLC method is unlikely to achieve the limit of detection at 1 ng/g level required by most of the countries that ban the drugs, and consequently not only more sensitive but also much costly liquid chromatography-mass spectrometry (LC-MS) is required for detecting CV or MG at such low level [19]. SERS technology has shown great potential for detecting trace amounts of analytes with a simpler sample extraction protocol and a shorter detection time, which makes it possible to rapidly screen and analyze the use of illegal drugs in aquatic products [20].

The objective of this study was to investigate the potential of applying Au-Ag core-shell bimetallic nanoparticles to the detection of banned triphenylmethane drugs in fish muscle. This study provides a basis for applying SERS technology to the identification of trace amounts of prohibited substances in aquatic food products, and the methodology could be extended to analyses of other hazardous chemicals in complex food matrices (such as fish and meat) with SERS technology.

MATERIALS AND METHODS

Synthesis of Au-Ag Core-Shell Nanoparticles

Au-Ag core-shell NPs were synthesized in solution via a seed-growth method [21]. In brief, sodium citrate (0.74 mL, 1% w/w) was added to the boiling solution of chloroauric acid (50 mL, 2×10^{-4} mol/L), and the mixture was stirred and boiled until the color became wine red, indicating that Au NPs were formed [22]. As-prepared Au colloids (3 mL) and L-ascorbic acid (0.4 mL, 0.1 mol/L) were added in a vial under continuous stirring. Then 0.3 mL, 0.6 mL, 0.9 mL, and 1.2 mL of silver nitrate (1×10^{-3} mol/L) were added dropwise (10 μL per addition) to this mixture to prepare four different Au-Ag core-shell NPs varying in the thickness of silver coating, respectively. The formed Au-Ag core-shell NPs were transferred into a conical flask with stopper and kept in refrigerator at 4°C before use.

Preparation of Standard Solutions

CV (≥90%, Sigma-Aldrich, USA) and MG (>99%, Sigma-Aldrich, USA) were dissolved in acetonitrile (HPLC reagent, Sigma, USA) aqueous solution (v/v = 1 : 1) to prepare a series of standard solutions (0.1, 1, 10, and 10^3 µg/L). The pH of the CV and MG standard solutions ranged from 5.3 to 5.8, depending on the concentration of the drugs. The absorbance bands of CV and MG were at 590 nm and 620 nm, respectively, based upon UV-Vis spectroscopic analysis (UV3000PC, MAPADA Instruments Ltd., Shanghai, China).

Fish Sample Pretreatment

Tilapia fillets (containing no CV or MG as confirmed with LC-MS) from Zhenye Aquatic and Cool Storage Ltd. (Guangdong, China) were used in this study. Tilapia fillets were homogenized through blending frozen tilapia filets with dry ice in a laboratory blender (HGBTWTS3, Waring Commercial, Torrington, CT, USA) at high speed for 5 minutes to achieve uniform fish samples and therefore minimize spectral variation due to the disparity of nontargeted components in fish tissue. The extraction and purification protocol for CV and MG in fish muscle were based upon a conventional method adopted by the US Food and Drug Administration (FDA) with slight modifications [23]. In brief, homogenized tilapia fillets spiked with CV or MG [0 (blank), 0.1, 1, and 10 ng/g] were blended with ammonium acetate buffer (Sinopharm Chemical Reagent Ltd., SCRC, Shanghai, China), hydroxylamine hydrochloride solution (ACS reagent, SCRC), and p-toluenesulfonic acid solution (ACS reagent, SCRC) and mixed well. Then, acetonitrile and alumina (chromatographic grade, SCRC) were added and the mixture was centrifuged, followed by the use of supernatants for liquid-liquid extraction with the addition of dichloromethane (ACS reagent, SCRC). The extract was evaporated to dry with a rotary evaporator (R206B, Shanghai SENCO Technology Ltd., Shanghai, China), dissolved into acetonitrile, and mixed with 2,3-dichloro-5,6-dicyano-1,4-benzoquinone (DDQ) solution (98%; J&K SCIENTIFIC, Logan, UT, USA). The solution was further purified with solid phase extraction through alumina cartridge (1 g, 3 mL; Supelco, Bellefonte, PA, USA) positioned on top of a propylsulfonic acid cartridge (500 mg, 3 mL; ANPEL Scientific Instrument Co., Ltd., Shanghai, China). Unlike the FDA method, no vacuum pump was used to speed up the flow rate during the solid phase extraction. The final elute was reconstituted to 5 mL with ammonium acetate and acetonitrile (v/v = 1 : 1).

SERS Measurement

SERS spectra were collected by using a Nicolet DXR microscopy Raman spectrometer (Thermo Fisher Scientific Inc., Waltham, MA, USA) with a 633 nm He-Ne laser source, 2 mW laser power, and 20x objective lens with a slit width of 50 cm^{-1}.

To collect an SERS spectrum, 10 µL Au-Ag core-shell NPs were deposited onto a microscope glass slide, and after evaporation of solvent, one drop of sample solution or fish extract was pipetted onto the Au-Ag core-shell substrate. SERS spectra were immediately acquired after evaporation of the solvent. The exposure time was 20 s for each scan and each spectrum was the average of 5 scans. For each treatment of standard solutions (10^3, 10, 1, and 0.1 µg/L) or fish fillets spiked with CV or MG [0 (blank), 0.1, 1, and 10 ng/g], 10 spectra from different locations of a substrate were collected, and triplicate analyses were conducted.

Characterizations of Nanoparticles

The optical properties of Au NPs used as seeds and Au-Ag core-shell NPs coated with different amounts of silver were analyzed with a UV-Vis absorbance spectroscopy (UV3000PC, MAPADA Instruments Ltd., Shanghai, China). General shapes of the Au and Au-Ag core-shell NPs were analyzed with a transmission electron microscopy (TEM; JEM-2100F, JEOL Ltd., Tokyo, Japan) and the core-shell structure of Au-Ag core-shell NPs was determined with a high-resolution TEM (HRTEM, JEM-2100F, JEOL Ltd., Tokyo, Japan). The average particle sizes of Au seeds and bimetallic NPs were calculated based upon TEM images of 50 particles.

CV standard solution (10 µg/L) was used as probe to understand the effect of the synthesized Au-Ag core-shell substrates on SERS enhancement. The SERS spectra of CV with the use of four different Au-Ag core-shell NPs as substrates were recorded, respectively, and the enhancement factors of these substrates were estimated and compared [24]. The substrate resulting in the highest enhancement for CV was selected for the further study on analysis of CV and MG standard solutions and fish fillets spiked with CV or MG.

To evaluate the stability of Au-Ag core-shell bimetallic nanosphere colloids, the selected Au-Ag core-shell NPs were stored at refrigerator temperature for up to 20 days. Sixteen spectra of CV standard solution (1 µg/L, 10 µg/L) with the use of the selected colloids as SERS substrates were acquired every other day. The Raman intensities at $1617\,cm^{-1}$ (one of the primary characteristic peaks for CV) were compared among spectra collected from different days to see whether there was significant difference ($\alpha=0.05$) with ANOVA analysis (Excel 2010; Microsoft Co., Redmond, WA, USA).

RESULTS AND DISCUSSION

UV-Vis Spectra of Au-Ag Core-Shell NPs with Different Sizes

The surface plasmon resonance (SPR) peak of metal NPs can be detected with UV-Vis absorbance spectroscopy. It is well known that the position and shape of SPR peak depend upon the shape, size, and composition of NPs [25], and all these factors ultimately affect the enhancement effect of SERS substrates. The reported SPR peaks of Au NPs ranged from 517 nm to 575 nm (particle diameter: 9–99 nm), while the peaks of Ag NPs ranged from 390 nm to 438 nm (particle diameter: 10–80 nm), and the SPR peak tended to shift to higher wavelength (red shift) with an increase in particle sizes for Au or Ag NPs [26–28]. The optical properties of Au-Ag core-shell NPs are more complex than those of Au or Ag NPs because of an interaction between Au and Ag. As shown in Figure 1, the change in the plasmon resonance of Au-Ag core-shell NPs during the silver coating process was clearly affected by the ratio of Au to Ag and the thickness of Ag shell. Before adding $AgNO_3$, Au NPs dispersion exhibited absorbance band at about 521 nm, and the shape of the SPR peak was symmetrical and narrow, indicating relatively good monodispersion of the Au NPs (diameter: 18±2 nm). With the addition of $AgNO_3$ solution and ascorbic acid,

Ag⁰ was produced and gradually coated on the surface of Au seeds as indicated by the appearance of characteristic peak of silver shells (376–400 nm) as well as the shifts of SPR peaks for both Ag shells and Au seeds (Figure 1). As the amount of $AgNO_3$ solution increased from 0.3 mL to 1.2 mL, the SPR peak of Ag shells red-shifted from 376 nm to 400 nm and the peak intensity also strengthened due to increase in the thickness of silver layer deposited onto the Au seeds [29], while the SPR peak for Au seeds showed opposite trend and was barely discernible when the amount of $AgNO_3$ solution reached up to 0.9 mL (particle diameter: 33±3 nm) and above. The overall SPR peaks of Ag shells for Au-Ag core-shell NPs appeared at relatively low wavelengths (376–400 nm) compared to those reported for Ag NPs (390–438 nm), which was mainly because the Ag shells (5–11 nm) were quite thin compared to the radiuses of the reported Ag NPs (5–40 nm) [27, 28].

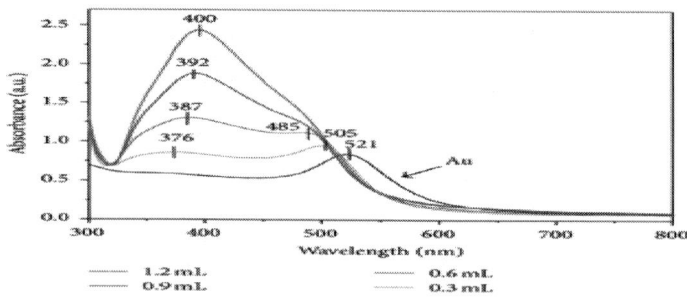

Figure 1: Changes of UV-Vis spectra of Au-Ag core-shell nanoparticles dispersion with the increased amounts of $AgNO_3$ used for synthesis.

Nanoparticles Size-Dependent SERS Enhancement Effect for CV

Figure 2 shows some representative SERS spectra of 10 µg/L CV standard solution acquired by using four Au-Ag core-shell substrates synthesized with different amounts of $AgNO_3$. The observed main characteristic peaks of CV at around 1617, 1367, 1173, 914, and 440 cm⁻¹ are assigned to the in-plane stretching of the ring–C–C, N–phenyl stretching, in-plane bending of the ring–C–H, ring skeletal vibration of radical orientation, and out-of-plane deformation vibrations of the phenyl–C⁺–

phenyl [30, 31]. As can be seen from the change in the intensity of CV characteristic peaks, an increasing enhancement effect was observed with the use of thicker silver-coated Au-Ag core-shell NPs when the amount of $AgNO_3$ used for silver coating was no more than 0.9 mL. The Au-Ag core-shell NPs could reach sufficiently intrinsic SERS activity of the growing Ag shells, which is much higher than Au NPs to generate strong electromagnetic enhancement for high SERS signals [32]. However, the enhancement effect decreased when the amount of $AgNO_3$ used for silver coating reached 1.2 mL. The enhancement factors for four Au-Ag core-shell substrates were calculated as 2.5×10^5, 8.0×10^5, 1.8×10^6, and 6.1×10^5, corresponding to 0.3, 0.6, 0.9, and 1.2 mL of $AgNO_3$ used for silver coating, respectively. Therefore, the Au-Ag core-shell NPs synthesized with 0.9 mL of $AgNO_3$ (particle diameter: 33 ± 3 nm) were selected for further analysis of CV and MG in fish muscle.

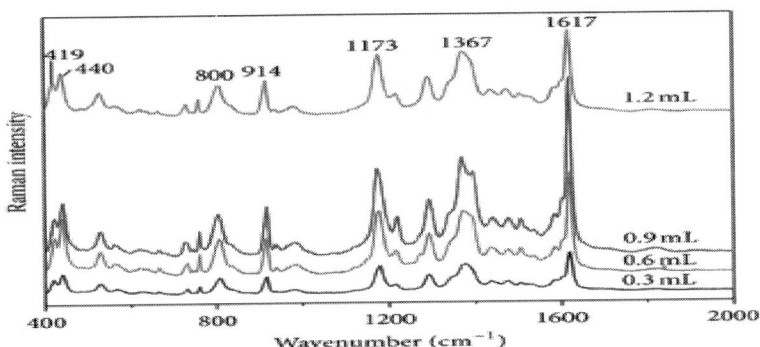

Figure 2: Representative SERS spectra of 10 μg/L crystal violet solution by using Au-Ag core-shell substrates synthesized with different amounts of $AgNO_3$.

Based on the calculation of the TEM images of 50 particles, the average particle size of Au seeds was 18 ± 2 nm in diameter, and the sizes of Au-Ag core-shell NPs were 28 ± 4nm, 30 ± 4 nm, 33 ± 3 nm, and 39 ± 3 nm, corresponding to 0.3, 0.6, 0.9, and 1.2 mL of $AgNO_3$ used for silver coating, respectively. The TEM images of the Au seeds and the 33 nm Au-Ag core-shell NPs showed that both NPs were relatively uniform, nearly spherical with narrow size distribution (relative standard deviation: Au NPs, 11%; 33 ± 3nm Au-Ag core-shell NPs, 9%) (Figures

3(a) and 3(b)). The average thickness of the Ag layer on the Au core is 7.5 nm for the selected Au-Ag core-shell NPs. In addition, no seed-size Au NPs or smaller Ag NPs were observed, suggesting that no other nucleation centers or separation from the Au-core layer occurred [6, 33]. The HRTEM image of Au-Ag core-shell NPs (Figure 3(c)) showed that some lateral moire fringes were generated which indicated the selective growing of Ag [34–36].

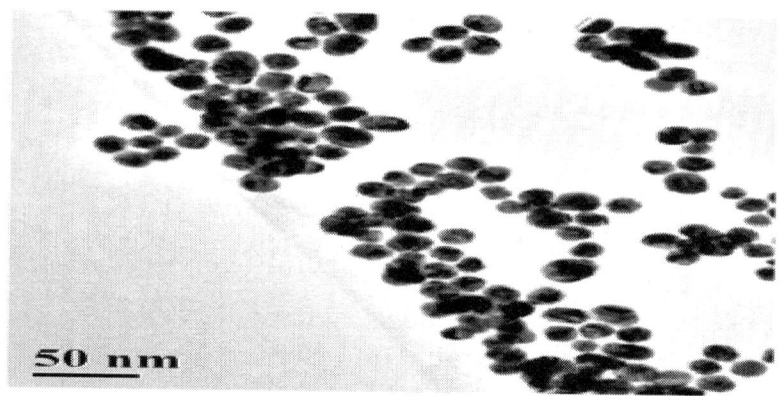

(a)

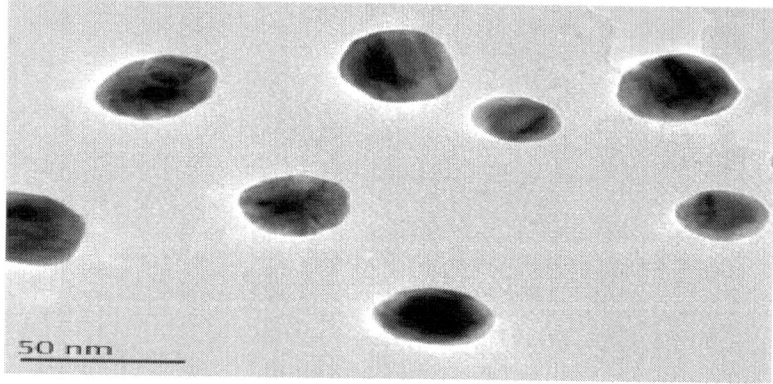

(b)

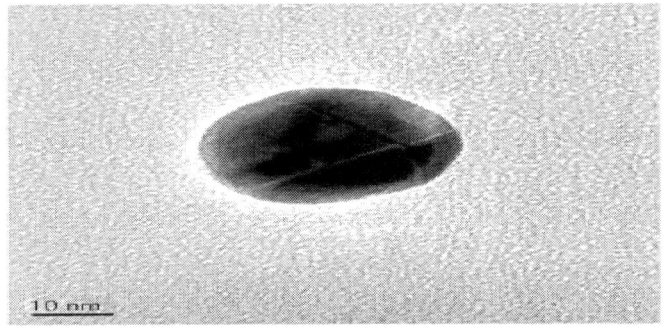

(c)

Figure 3: TEM images of (a) Au nanoparticles and (b) Au-Ag core-shell nanoparticles (0.9 mL AgNO$_3$ used) and (c) HRTEM image of Au-Ag core-shell nanoparticles (0.9 mL AgNO$_3$ used).

During 20-day storage at refrigerator temperature, with the selected 33 nm Au-Ag core-shell NPs as substrates to acquire SERS spectra of CV standard solution (1 µg/L, 10 µg/L), there was no significant change in the Raman intensities at 1617 cm^{-1} for both 1 µg/L and 10 µg/L solutions, indicating that the colloids were stable and no obvious change occurred in 20 days.

Analysis of CV and MG with Au-Ag Core-Shell SERS Substrates

A series of CV and MG standard solution (0.1, 1, 10, and 10^3 µg/L) was used to evaluate the SERS activity of the selected 33 nm Au-Ag core-shell substrates. As can be seen from Figure 4, there was almost no SERS signal for blank substrates, whereas the intensity of the characteristic bands was drastically enhanced as the concentration of CV or MG solution increased from 0.1 µg/L to 10 µg/L. Even at the concentration level as low as 0.1 µg/L, the major characteristic bands at around 440, 914, 1173, 1367 cm^{-1}, and 1617 cm^{-1} could still be identified for both CV and MG. SERS spectra of MG were similar to those of CV because of their similar molecular structures, and only changes in the relative intensity among peaks were observed, such that the relative intensity

of the peak at around 1217 cm⁻¹ for MG was stronger than that for CV
[20, 37].

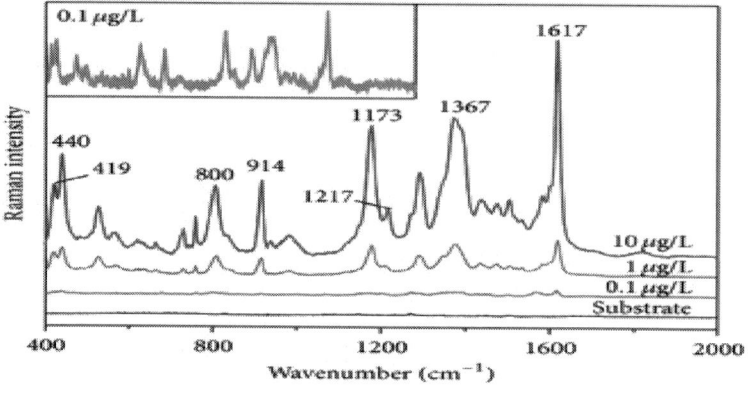

(a)

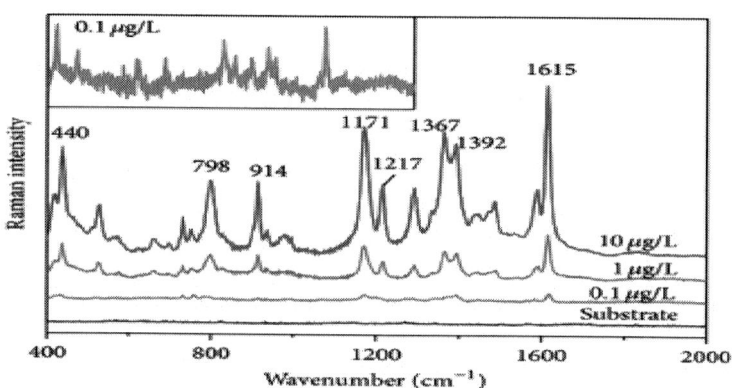

(b)

Figure 4: SERS spectra of (a) crystal violet standard solution and (b) malachite
green standard solution deposited on Au-Ag core-shell nanoparticles.

The optimal Au-Ag core-shell substrates were further used to
detect CV and MG residues in fish samples with conventional sample

preparation method. Figure 5 presents SERS spectra of CV and MG extracts [0 (blank), 0.1, 1, and 10 ng/g] adsorbed on Au-Ag core-shell substrates. There was no SERS signal detected for the blank extracts, which indicates that sample extract was relatively clean and the presence of some nontargeted compounds did not have obvious Raman scattering signals that may interfere with the analysis of CV or MG. The main spectral characteristic peaks of fish extracts with different levels of CV or MG were basically consistent with those of the standard solution, showing similar intensity and no obvious interference by other components from fish fillets. In addition, the major characteristic bands were discernable at a level as low as 0.1 ng/g for both CV and MG in contaminated fish samples, which were similar to the results for CV or MG standard solutions as discussed earlier. A high recovery rate (85.9–93.9%) was obtained by using the conventional methods for CV and MG extraction [38]. This high recovery rate of CV and MG together with little or no interference of other components in CV and MG extracts accounted for the similar SERS results for analysis of CV or MG in fish extracts and in standard solutions.

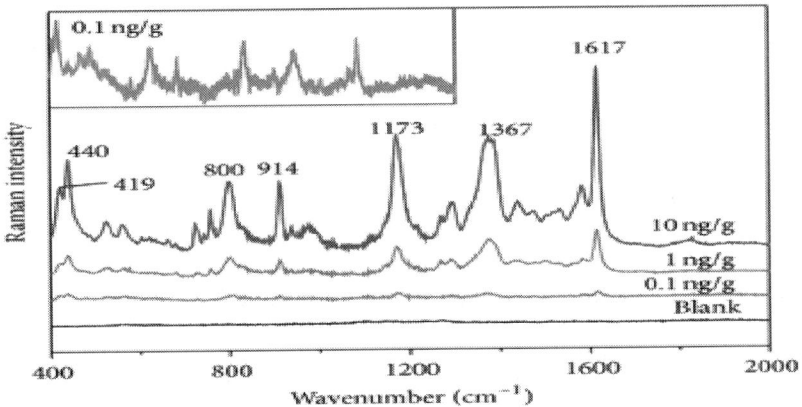

(a)

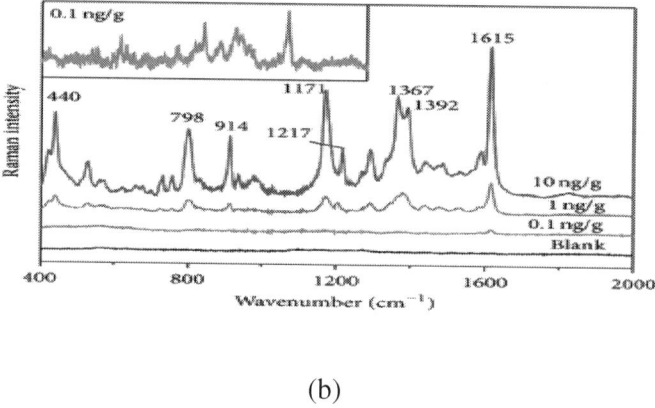

(b)

Figure 5: SERS spectra of (a) crystal violet extracts and (b) malachite green extracts from fish muscle.

Obtaining reproducible spectra has been always a challenge for applying SERS technology as an analytical tool, since the intensities of SERS signals were greatly affected by the surface morphology of substrates, location, and direction of the targeted molecules adsorbed on a substrate and many other factors. To achieve reproducible results, ten spectra were acquired from different locations of a substrate each time during SERS collection, and the average of these ten spectra was used as the final spectra. Figure 6 exhibits the average spectra from triplicate analyses for CV and MG of both standard solutions (1 µg/L) and fish extracts (1 ng/g), indicating that relatively repeatable SERS spectra could be achieved with the selected Au-Ag core-shell NPs as substrates.

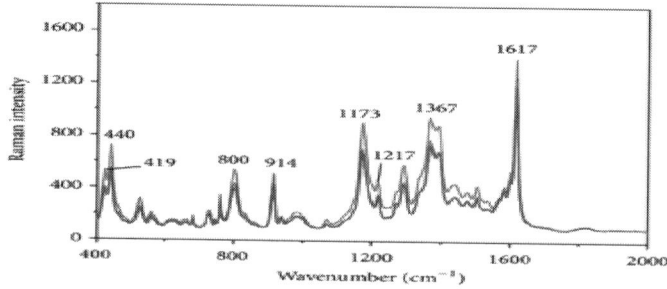

(a)

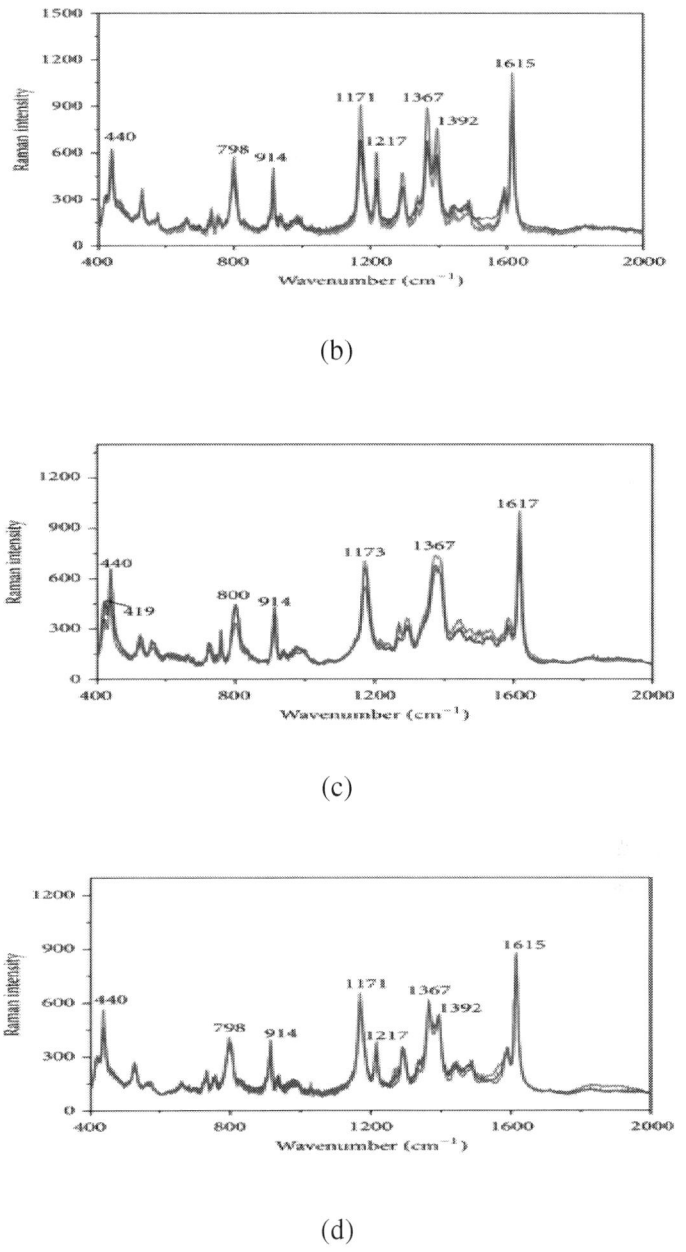

(b)

(c)

(d)

Figure 6: Average SERS spectra (n=10) of (a) CV standard solutions (1 µg/L), (b) MG standard solutions (1 µg/L), (c) CV fish extract (1 ng/g), and (d) MG fish extract (1 ng/g) from triplicate analyses.

Most of CV or MG dyes are metabolized to leucocrystal violet or leucomalachite green in fish [19]. The conventional sample preparation method applied in this study to extract CV or MG from fish fillet involved the use of DDQ to convert metabolites leucocrystal violet or leucomalachite green to CV or MG, and the final level of fish drug detected was actually the sum of CV (or MG) and its metabolite. Therefore, although only CV or MG was tested, the study results could be applied to analyze CV (or MG) and its metabolite in fish muscle.

CONCLUSIONS

Silver-coated gold bimetallic NPs were synthesized and successfully applied to detect as low as 0.1 ng/g of CV or MG in fish muscle, which made great improvement over our previous studies using commercial SERS substrates [20] and also could meet the most restricted regulatory requirements for the 1 ng/g limit of detection for crystal violet or malachite green in fish muscle. This interdisciplinary study combined knowledge in SERS and analytical chemistry to solve a real world food safety problem, which helps to fill the gap between theory and application aspects of SERS-related research. Although there were reports on using SERS for nondestructive analyses of chemical hazards in food, such as pesticide in apples [35], since CV or MG residues are inside fish tissues, it is impossible to apply SERS technology to analyze these drugs without sample preparation involving extraction and purification. In this study, a general sample preparation method for chromatography-based analyses was adopted; the information gained from the study provides a basis for further explorations on simplifying sample preparation methods as well as for expanding SERS methods with bimetallic nanospheres for analysis of other harmful chemical substances, such as pesticides and antibiotics in food systems.

ACKNOWLEDGMENTS

This research was supported by the National Natural Science Foundation of China (61250002 and 31250006).

REFERENCES

1. D. L. Jeanmaire and R. P. van Duyne, "Surface Raman spectroelectrochemistry—part I: heterocyclic, aromatic, and aliphatic amines adsorbed on the anodized silver electrode," Journal of Electroanalytical Chemistry and Interfacial Electrochemistry, vol. 84, no. 1, pp. 1–20, 1977.

2. C. L. Haynes, A. D. McFarland, and R. P. van Duyne, "Surface-enhanced Raman spectroscopy,"Analytical Chemistry A, vol. 77, no. 17, pp. 338–346, 2005.

3. E. Petryayeva and U. J. Krull, "Localized surface plasmon resonance: nanostructures, bioassays and biosensing—a review," Analytica Chimica Acta, vol. 706, no. 1, pp. 8–24, 2011.

4. D. Cialla, A. März, R. Böhme et al., "Surface-enhanced Raman spectroscopy (SERS): progress and trends," Analytical and Bioanalytical Chemistry, vol. 403, no. 1, pp. 27–54, 2012.

5. S. Nie, "Probing single molecules and single nanoparticles by surface-enhanced Raman scattering,"Science, vol. 275, no. 5303, pp. 1102–1106, 1997.

6. M. Mandal, N. R. Jana, S. Kundu, S. K. Ghosh, M. Panigrahi, and T. Pal, "Synthesis of Aucore-Agshell type bimetallic nanoparticles for single molecule detection in solution by SERS method," Journal of Nanoparticle Research, vol. 6, no. 1, pp. 53–61, 2004.

7. M. V. Cañamares, C. Chenal, R. L. Birke, and J. R. Lombardi, "DFT, SERS, and single-molecule SERS of crystal violet," The Journal of Physical Chemistry C, vol. 112, no. 51, pp. 20295–20300, 2008.

8. L. Lu, H. Wang, Y. Zhou et al., "Seed-mediated growth of large, monodisperse core-shell gold-silver nanoparticles with Ag-like optical properties," Chemical Communications, no. 2, pp. 144–145, 2002.

9. F. Zhai, Y. Huang, C. Li, X. Wang, and K. Lai, "Rapid determination of ractopamine in swine urine using surface-enhanced raman spectroscopy," Journal of Agricultural and Food Chemistry, vol. 59, no. 18, pp. 10023–10027, 2011.

10. Y. Li, H. Su, K. S. Wong, and X.-Y. Li, "Surface-enhanced raman spectroscopy on two-dimensional networks of gold nanoparticle-nanocavity dual structures supported on dielectric nanosieves,"

The Journal of Physical Chemistry C, vol. 114, no. 23, pp. 10463–10477, 2010.

11. N. R. Jana, "Silver coated gold nanoparticles as new surface enhanced Raman substrate at low analyte concentration," Analyst, vol. 128, no. 7, pp. 954–956, 2003.

12. Y. Cui, B. Ren, J.-L. Yao, R.-A. Gu, and Z.-Q. Tian, "Synthesis of AgcoreAushell bimetallic nanoparticles for immunoassay based on surface-enhanced Raman spectroscopy," The Journal of Physical Chemistry B, vol. 110, no. 9, pp. 4002–4006, 2006.

13. L. Guerrini, J. V. Garcia-Ramos, C. Domingo, and S. Sanchez-Cortes, "Ultrathin silver-coated gold nanoparticles as suitable substrate for surface-enhanced raman scattering," Journal of Raman Spectroscopy, vol. 41, no. 5, pp. 508–515, 2010.

14. S. Srivastava, R. Sinha, and D. Roy, "Toxicological effects of malachite green," Aquatic Toxicology, vol. 66, no. 3, pp. 319–329, 2004.

15. N. A. Littlefield, B. N. Blackwell, C. C. Hewitt, and D. W. Gaylor, "A novel surface-enhanced Raman scattering sensor to detect prohibited colorants in food by graphene/silver nanocomposite,"Fundamental and Applied Toxicology, vol. 5, pp. 902–912, 1985.

16. A. Stammati, C. Nebbia, I. de Angelis et al., "Effects of malachite green (MG) and its major metabolite, leucomalachite green (LMG), in two human cell lines," Toxicology in Vitro, vol. 19, no. 7, pp. 853–858, 2005.

17. RAFFS, Subject: crystal violet or malachite green. Notification between 01/01/2003 to 31/12/2012, 2013,https://webgate.ec.europa.eu/rasff-window/portal.

18. A. A. Bergwerff and P. Scherpenisse, "Determination of residues of malachite green in aquatic animals,"Journal of Chromatography B: Analytical Technologies in the Biomedical and Life Sciences, vol. 788, no. 2, pp. 351–359, 2003.

19. W. C. Andersen, S. B. Turnipseed, C. M. Karbiwnyk et al., "Multiresidue method for the triphenylmethane dyes in fish: malachite green, crystal (gentian) violet, and brilliant green," Analytica Chimica Acta, vol. 637, no. 1-2, pp. 279–289, 2009.

20. Y. Zhang, K. Lai, J. Zhou, X. Wang, B. A. Rasco, and Y. Huang, "A novel approach to determine leucomalachite green and malachite green in fish fillets with surface-enhanced Raman spectroscopy (SERS) and multivariate analyses," Journal of Raman Spectroscopy, pp. 1208–1213, 2012.

21. T. Y. Olson, A. M. Schwartzberg, C. A. Orme, C. E. Talley, B. O'Conneull, and J. Z. Zhang, "Hollow gold-silver double-shell nanospheres: structure, optical absorption, and surface-enhanced Raman scattering," The Journal of Physical Chemistry C, vol. 112, no. 16, pp. 6319–6329, 2008.

22. G. Frens, "Controlled nucleation for the regulation of the particle size in monodisperse gold suspensions," Nature Physical Science, vol. 241, pp. 20–22, 1973.

23. W. C. Andersen, S. B. Turnipseed, and J. E. Roybal, "FDA laboratory information Bulletin," LIB 4363, 2005, Laboratory Information Bulletin U.S. Food and Drug Administration, November 2005.

24. C. Ruan, W. Wang, and B. Gu, "Single-molecule detection of thionine on aggregated gold nanoparticles by surface enhanced Raman scattering," Journal of Raman Spectroscopy, vol. 38, no. 5, pp. 568–573, 2007.

25. S. He, J. Yao, P. Jiang et al., "Formation of silver nanoparticles and self-assembled two-dimensional ordered superlattice," Langmuir, vol. 17, no. 5, pp. 1571–1575, 2001.

26. S. Link and M. A. El-Sayed, "Size and temperature dependence of the plasmon absorption of colloidal gold nanoparticles," The Journal of Physical Chemistry B, vol. 103, no. 21, pp. 4212–4217, 1999.

27. S. D. Solomon, M. Bahadory, A. V. Jeyarajasingam, S. A. Rutkowsky, C. Boritz, and L. Mulfinger, "Synthesis and study of silver nanoparticles," Journal of Chemical Education, vol. 84, no. 2, pp. 322–325, 2007.

28. C. S. Seney, B. M. Gutzman, and R. H. Goddard, "Correlation of size and surface-enhanced raman scattering activity of optical and spectroscopic properties for silver nanoparticles," The Journal of Physical Chemistry C, vol. 113, no. 1, pp. 74–80, 2009.

29. Y. Yang, J. Shi, G. Kawamura, and M. Nogami, "Preparation of Au-Ag, Ag-Au core-shell bimetallic nanoparticles for surface-

enhanced Raman scattering," Scripta Materialia, vol. 58, no. 10, pp. 862–865, 2008.

30. E. J. Liang, X. L. Ye, and W. Kiefer, "Surface-enhanced Raman spectroscopy of crystal violet in the presence of halide and halate ions with near-infrared wavelength excitation," The Journal of Physical Chemistry A, vol. 101, no. 40, pp. 7330–7335, 1997.

31. K. R. Strehle, D. Cialla, P. Rösch, T. Henkel, M. Köhler, and J. Popp, "A reproducible surface-enhanced Raman spectroscopy approach. Online SERS measurements in a segmented microfluidic system,"Analytical Chemistry, vol. 79, no. 4, pp. 1542–1547, 2007.

32. Z. Yi, X. Xu, X. Li et al., "Facile preparation of Au/Ag bimetallic hollow nanospheres and its application in surface-enhanced Raman scattering," Applied Surface Science, vol. 258, no. 1, pp. 212–217, 2011.

33. F.-K. Liu, M.-H. Tsai, Y.-C. Hsu, and T.-C. Chu, "Analytical separation of Au/Ag core/shell nanoparticles by capillary electrophoresis," Journal of Chromatography A, vol. 1133, no. 1-2, pp. 340–346, 2006.

34. C. Langlois, D. Alloyeau, Y. Le Bouar et al., "Growth and structural properties of CuAg and CoPt bimetallic nanoparticles," Faraday Discussions, vol. 138, pp. 375–391, 2008.

35. B. Liu, G. Han, Z. Zhang et al., "Shell thickness-dependent Raman enhancement for rapid identification and detection of pesticide residues at fruit peels," Analytical Chemistry, vol. 84, no. 1, pp. 255–261, 2012.·

36. R. Juluri, A. Rath, A. Ghosh et al., "Coherently embedded Ag nanostructures in Si: 3D imaging and their application to SERS," Scientific Reports, 2014.

37. L. He, N.-J. Kim, H. Li, Z. Hu, and M. Lin, "Use of a fractal-like gold nanostructure in surface-enhanced Raman spectroscopy for detection of selected food contaminants," Journal of Agricultural and Food Chemistry, vol. 56, no. 21, pp. 9843–9847, 2008.

38. W. C. Andersen, S. B. Turnipseed, and J. E. Roybal, "Quantitative and confirmatory analyses of malachite green and leucomalachite green residues in fish and shrimp," Journal of Agricultural and Food Chemistry, vol. 54, no. 13, pp. 4517–4523, 2006.

Citations

CHAPTER 1

A. V. Yakovlev and O. Yu. Golubeva, "Synthesis Optimisation of Lyso-zyme Monolayer-Coated Silver Nanoparticles in Aqueous Solution," Journal of Nanomaterials, vol. 2014, Article ID 460605, 8 pages, 2014. doi:10.1155/2014/460605.

CHAPTER 2

Mustafa Culha, Brian Cullum, Nickolay Lavrik, and Charles K. Klu-tse, "Surface-Enhanced Raman Scattering as an Emerging Character-ization and Detection Technique," Journal of Nanotechnology, vol. 2012, Article ID 971380, 15 pages, 2012. doi:10.1155/2012/971380.

CHAPTER 3

Hua Qi, R. W. Rendell, O. J. Glembocki, and S. M. Prokes, "Polarization Dependence of Surface Enhanced Raman Scattering on a Single Dielectric Nanowire," Journal of Nanomaterials, vol. 2012, Article ID 946868, 9 pages, 2012. doi:10.1155/2012/946868.

CHAPTER 4

Seongmin Hong and Xiao Li, "Optimal Size of Gold Nanoparticles for Surface-Enhanced Raman Spectroscopy under Different Conditions," Journal of Nanomaterials, vol. 2013, Article ID 790323, 9 pages, 2013 doi:10.1155/2013/790323.

CHAPTER 5

J. Betzabe González-Campos, Evgen Prokhorov, Isaac C. Sanchez, et al., "Molecular Dynamics Analysis of PVA-AgnP Composites by Dielectric Spectroscopy," Journal of Nanomaterials, vol. 2012, Article ID 925750, 11 pages, 2012. doi:10.1155/2012/925750.

CHAPTER 6

Ujjwala Gaware, Vaishali Kamble, and Balaprasad Ankamwar, "Ecofriendly Synthesis of Anisotropic Gold Nanoparticles: A Potential Candidate of SERS Studies," International Journal of Electrochemistry, vol. 2012, Article ID 276246, 6 pages, 2012. doi:10.1155/2012/276246.

CHAPTER 7

Emilia Tomaszewska, Katarzyna Soliwoda, Kinga Kadziola, et al., "Detection Limits of DLS and UV-Vis Spectroscopy in Characterization of Polydisperse Nanoparticles Colloids," Journal of Nanomaterials, vol. 2013, Article ID 313081, 10 pages, 2013. doi:10.1155/2013/313081.

CHAPTER 8

Norfazila Mohd Sultan and Mohd Rafie Johan, "Synthesis and Ultraviolet Visible Spectroscopy Studies of Chitosan Capped Gold Nanoparticles and Their Reactions with Analytes," The Scientific World Journal, vol. 2014, Article ID 184604, 7 pages, 2014. doi:10.1155/2014/184604.

CHAPTER 9

Lu Pei, Yiqun Huang, Chunying Li, Yuanyuan Zhang, Barbara A. Rasco, and Keqiang Lai, "Detection of Triphenylmethane Drugs in Fish Muscle by Surface-Enhanced Raman Spectroscopy Coupled with Au-Ag Core-Shell Nanoparticles," Journal of Nanomaterials, vol. 2014, Article ID 730915, 8 pages, 2014. doi:10.1155/2014/730915.

Index